gene discovery lab

Philip N. Benfey
New York University

BROOKS/COLE

THOMSON LEARNING

Australia • Canada • Mexico • Singapore • Spain • United Kingdom • United States

Printed in Canada

10   9   8   7   6   5   4   3   2   1

ISBN: 0-534-37717-3

Introduction to Techniques

<u>PURPOSE</u>

The exercises in this module are designed to familiarize you with the equipment and techniques used in the experimental protocols of the GDL modules.

<u>TABLE OF CONTENTS</u>

<u>Tool Bar</u>
<u>Introduction</u>
<u>Exercises that prepare for each module</u>
<u>Exercise 1: Using the pipettor.</u>
<u>Exercise 2: Selecting tubes to fill a rack.</u>
<u>Exercise 3: Moving tubes</u>
<u>Exercise 4: Designing oligonucleotide primers</u>
<u>Exercise 5: Putting filters into roller bottles</u>
<u>Exercise 6: Using the heat block</u>
<u>Exercise 7: Using the hybridization oven</u>
<u>Exercise 8: Loading a gel</u>
<u>Exercise 9: Using the UV crosslinking oven</u>
<u>Exercise 10: Using the PCR Machine</u>

<u>TOOLBAR</u>

A toolbar at the bottom of the screen provides useful features of GDL:
SHOW PROTOCOL brings up this window. It can be moved and resized.
SHOW NOTEBOOK brings up your notebook. It can be moved and resized.
QUIZ brings up a self-test of the knowledge gained from the experiments.
SAVE allows you to save a session.
MODULES allows you to go to another module in GDL.
HELP provides answers to general questions about how the laboratory works.
GLOSSARY provides definitions of terms used in GDL.
QUIT exits you from GDL.
BACK moves you to the previous screen.
SCREEN NUMBER tells you which screen you are on.
NEXT moves you to the next screen.

<u>INTRODUCTION</u>

In this module, you can perform many of the tasks you will need in order to complete the experiments in the other modules. The exercises are designed to familiarize you with the processes and equipment found in the laboratory before you use them to obtain an experimental result. Each of the exercises is self-contained and can be repeated as often as you like. You can work through all of the exercises before going to the other modules, or you can do only those exercises that are relevant to the module you are working on.

<u>cDNA Cloning</u>
<u>Exercise 1: Using the pipettor.</u>
<u>Exercise 2: Selecting tubes to fill a rack.</u>
<u>Exercise 3: Moving tubes.</u>
<u>Exercise 4: Designing oligonucleotide primers.</u>
<u>Exercise 5: Putting filters into roller bottles.</u>
<u>Exercise 6: Using the heat block.</u>
<u>Exercise 7: Using the hybridization oven.</u>

<u>Restriction Mapping</u>
<u>Exercise 1: Using the pipettor.</u>
<u>Exercise 2: Selecting tubes to fill a rack.</u>
<u>Exercise 3: Moving tubes.</u>

Exercise 6: Using the heat block.
Exercise 8: Loading a gel.

Southern Blot
Exercise 1: Using the pipettor.
Exercise 2: Selecting tubes to fill a rack.
Exercise 3: Moving tubes.
Exercise 5: Putting filters into roller bottles.
Exercise 6: Using the heat block.
Exercise 7: Using the hybridization oven.
Exercise 9: Using the UV crosslinking oven.

Northern Blot
Exercise 1: Using the pipettor.
Exercise 2: Selecting tubes to fill a rack.
Exercise 3: Moving tubes.
Exercise 5: Putting filters into roller bottles.
Exercise 6: Using the heat block.
Exercise 7: Using the hybridization oven.
Exercise 9: Using the UV crosslinking oven.

Subcloning
Exercise 1: Using the pipettor.
Exercise 2: Selecting tubes to fill a rack.
Exercise 3: Moving tubes.
Exercise 6: Using the heat block.

DNA Sequencing
Exercise 1: Using the pipettor.
Exercise 2: Selecting tubes to fill a rack.
Exercise 3: Moving tubes.
Exercise 4: Designing oligonucleotide primers.
Exercise 6: Using the heat block.
Exercise 8: Loading a gel.

PCR
Exercise 1: Using the pipettor.
Exercise 2: Selecting tubes to fill a rack.
Exercise 3: Moving tubes.
Exercise 4: Designing oligonucleotide primers.
Exercise 6: Using the heat block.
Exercise 8: Loading a gel.
Exercise 10: Using the PCR Machine.

Protein Expression
Exercise 1: Using the pipettor.
Exercise 2: Selecting tubes to fill a rack.
Exercise 3: Moving tubes.
Exercise 6: Using the heat block.
Exercise 8: Loading a gel.

Exercise 1: Using the pipettor.

1. Drag the automatic pipettor from its rack. A new disposable tip appears automatically. The tip will be automatically discarded after each use.

2. Click on the dark rectangular window on the pipettor to set the volume.

3. Select the volume by adjusting the slider using the sliding needle or by using the arrow buttons for fine control.

4. Set the volume to 2 microliters.

5. ' Click OK or anywhere outside the slider panel to dismiss it.

6. Drag the pipettor to the enzyme cooler (the round container with tubes in it).

7. When the tip of the pipettor is over a tube, the contents of the tube will be visible. This tells you which tube the pipettor tip is in.

8. When you release the mouse, if the volume is set within the correct limits, the reagent will be drawn up into the pipettor tip. If not, you will get an error message and will need to reset the volume.

9. Draw up liquid from one of the tubes.

10. Drag the pipettor to the rack with tubes in it.

11. When the pipettor tip is over a tube, release the mouse. What was in the pipettor tip will now be added to that tube.

12. If you draw up something you don't want to use, drag the pipettor to the trash can, on the left and below the bench.

13. Release the pipettor over the trash and the current tip will be discarded, allowing you to start over with a new tip.

14. Adjust the volume of the pipettor from 2 microliters to 5 microliters, and transfer liquid from the tubes in the enzyme cooler to a tube in the rack.

15. Adjust the volume to 0.5 microliters, and try drawing up liquid to see what happens when the volume is set incorrectly.

Exercise 2: Selecting tubes to fill a rack.

1. Click NEXT on the toolbar at the bottom of the screen to move down the bench.

2. Click on the arrow on the front of the empty DNA tube rack .

3. On the menu that appears, hold down the mouse button as you move down the menu.

4. To select something from the menu, release the mouse button when the item you want is highlighted. A tube will appear in the rack, containing the selected item.

5. Place the cursor over a tube to reveal its contents.

6. To get another tube, click the arrow again and the menu will reappear.

7. Select tubes until you fill the rack.

8. To get rid of a tube, drag it to the trash.

Exercise 3: Moving tubes.

1. Click NEXT at the bottom of the screen to move down the bench.

2. To move a tube, place the cursor over the tube, depress the mouse button, and drag the tube to where you want it to go.

3. Drag a tube from the rack to the heat block.

4. To put the tube into a hole in the heat block, position the tube over the heat block, and release the mouse.

5. ' The tube will automatically go into the hole in the heat block.

6. Move all of the tubes from the rack to the heat block.

Exercise 4: <u>Designing oligonucleotide primers,</u>

1. Click NEXT to move down the bench.

2. Oligonucleotide primers are used to screen cDNA libraries, to sequence DNA, and for PCR amplification. An oligonucleotide is a short stretch of DNA whose sequence depends on the intended use. Each module explains how to choose the correct sequence .

3. Click the arrow on the Primer rack. On the menu that appears, select "Create new primer tube."

4. In the dialog box, use the four yellow keys (A, T, C, or G) to enter the nucleotides for your primer in the 5′ to 3′ direction. For places where you want to have more than one base (for example, in degenerate primers for probing cDNA libraries) use "/" to separate nucleotides at the same position.

5. Enter any sequence of nucleotides to get a feel for how it works. Your primer should be 20 bases long.

6. If you make a mistake, click the BACK button in the dialog box to erase that base.

Exercise 5: <u>Putting filters into roller bottles.</u>

1. Click NEXT to move down the bench.

2. In this exercise, you will use forceps to put filters into a roller bottle. To use the forceps, drag them toward a filter.

3. Using the forceps, drag one of the round white nylon filters to the roller bottle. When the tip of the forceps is over a filter, the forceps will close on the filter, and you can drag the filter..

4. When the filter is over the mouth of the roller bottle, the roller bottle will tilt and the filter will enter the roller bottle. Release the mouse to release the forceps from the filter.

5. Use the forceps to put all four filters into the roller bottle.

6. Put the cap on the roller bottle by dragging the cap from the bench and releasing it over the roller bottle. It will automatically drop onto the roller bottle.

Exercise 6: <u>Using the heat block.</u>

1. Click NEXT to move down the bench.

2. Turn on the heat block by clicking the ON/OFF switch.

3. Click the control knob on the heater to set the temperature.

4. On the slider that appears, set the temperature for 37 degrees C by moving the sliding needle or by using the arrows for fine control.

5. To start an incubation, click on the timer. Set the timer slider to 1 hour.

6. If you need to adjust anything, click outside the timer slider to dismiss it without starting the timer.

7. Click the GO button on the timer to start the timer and the incubation. If you click GO and you have forgotten to turn on the heat block, a message will appear to remind you.

8. The timer counts down the time and rings when the incubation is completed. In GDL, because time is speeded up, this will only take a few seconds.

Exercise 7: <u>Using the hybridization oven</u>.

1. Click NEXT to move down the bench.

2. Drag the roller bottle to the hybridization oven.

3. When the roller bottle approaches the door of the hybridization oven, the door will automatically open. Release the mouse to allow the roller bottle to drop into place in its holder.

4. Click on the panel to the left of the oven door to set the hybridization temperature. Use the slider to set the temperature to 42 degrees C.

5. To start a hybridization, click on the timer. Set the timer slider to 2 hours. If you need to adjust anything, click outside the timer slider to dismiss it without starting the timer.

6. Click GO on the timer to start the timer and the hybridization.

7. The timer counts down the time and rings when the hybridization is completed. In GDL, because time is speeded up, this will only take a few seconds.

8. After the hybridization is completed, click on the oven door to open it. The roller bottle will be brought back to the bench.

Exercise 8: <u>Loading a gel</u>.

1. Click NEXT to move down the bench.

2. In this exercise, you will practice loading a gel. You will perform the exercise with an agarose gel, but essentially the same procedure is used in loading a sequencing gel or a polyacrylamide gel.

3. Use the pipettor to draw up 10 microliters of liquid from one of the tubes in the rack. If you do not know how to use the pipettor, do the exercise <u>Using the pipettor</u>.

4. Drag the pipettor so that its tip is just above one of the wells (the small indentations) in the gel, which is in the gel box on the bench.

5. When the pipettor tip is over the selected well, release its contents into the well by releasing the mouse.

6. Load all the wells with the liquid in the tubes.

Exercise 9: <u>Using the UV cross-linking oven</u>.

1. Click NEXT to move down the bench.

2. In this exercise, you can practice putting a nylon filter (with DNA or RNA on it) into the UV cross-linking oven. Click on the door of the UV cross-linking oven to open it.

3. Drag the forceps to the white filter on the bench.

4. When the tip of the forceps is over a filter, release the mouse and the forceps will close on the filter. Drag the filter with the forceps to the UV cross-linking oven.

5. When the filter is inside the oven, release the mouse to release the filter from the forceps.

6. Click on the door to close it.

7. Click on the keypad to the right of the oven door to set the time of UV exposure. Move the slider to 10 seconds and click the GO button.

8. Click the ON/OFF switch to turn on the UV cross-linking oven.

9. When the UV cross-linking oven is turned on, a new window appears to allow you to see the filter illuminated by the UV light.

10. In the Southern Blot and Northern Blot modules, you can click on the film icon in the window to take a picture of the filter illuminated by UV light, which will appear in your notebook.

11. Click in the upper left-hand box on the window to dismiss it.

12. Click on the oven door to open it.

13. Drag the filter out with the forceps.

Exercise 10: Using the PCR Machine.

1. Click NEXT to move down the bench.

2. There is a rack of tubes on the bench ready to be put into the PCR machine. Drag the tubes one by one into the rack in the PCR machine.

3. Click on the cover to close it.

4. Click on the front panel of the PCR machine to get the set-up menu.

5. Set each of the parameters by releasing the mouse when the parameter is highlighted on the menu. Use the slider that appears to set the temperature, time, or number of cycles.

6. Set the annealing temperature to 60 degrees C. Set the time for the annealing step to 30 seconds.

7. Set the elongation temperature to 72 degrees C. Set the elongation time to 40 seconds.

8. Set the denaturation temperature to 94 degrees C. Set the denaturation time to 20 seconds.

9. Set the number of cycles to 30.

10. When all the steps have been programmed, select Start PCR from the menu.

11. If you have programmed it correctly, the PCR machine will indicate which cycle it is running.

cDNA cloning

OBJECTIVES

The procedures in this lab will enable you to:

- **Design an oligonucleotide primer that corresponds to the amino acid sequence of a protein.**
- **Hybridize the primer to a cDNA library.**
- **Identify the hybridizing clones among the background spots.**

PURPOSE

Screening cDNA libraries is a way to identify genes when the protein sequence is known.

You can use cDNA cloning to identify the gene that has been mutated in patients with the diseases under investigation.

To get started immediately, go to Task 1.

TABLE OF CONTENTS

Background
Introduction
Conceptual and technical overview
Task 1: Design primers to probe the cDNA library.
Task 2: Set up labeling reaction.
Task 3: Incubate the labeling reaction.
Task 4: Select cDNA libraries.
Task 5: Set up the prehybridization.
Task 6: Add probe to the roller bottle.
Task 7: Wash the filters.
Task 8: Place filters in the X-ray cassette.
Task 9: Interpret the results.
Troubleshooting
Equipment and reagent information

BACKGROUND

If you know where a gene is expressed then you can use a cDNA library to try to isolate the gene. To make a cDNA library, messenger RNA is isolated from a particular tissue or cell type, then reverse transcriptase is used to make a DNA copy of the RNA. If an RNA is abundant in the tissue from which the cDNA library is made, then it will be represented in a high proportion of the cDNA clones. Because cDNA libraries are usually made from messenger RNA, they do not contain introns. Thus, they can be used in bacteria to express proteins which are unable to remove introns.

To isolate a gene from a cDNA library, a probe is needed. Probes are pieces of DNA or RNA that are homologous to at least a portion of the gene that you wish to isolate. This means that the probe must be able to base-pair with the cDNA clone according to the base-pairing rules of A pairing with T and C pairing with G. In cases where you wish to identify a gene that encodes a particular protein (such as beta-globin or p53) you can use the amino acid sequence of the protein to "design" a probe. Because there is degeneracy in the genetic code for many amino acids there are more than one possible codon.

APPLICATIONS

Isolation of the human beta-globin gene is an excellent example of what is known as "reverse genetics." Starting from the amino acid sequence of the beta-globin protein, scientists made degenerate probes and were able to isolate a beta-globin cDNA. A portion of the cDNA clone was then used to isolate a genomic clone from a genomic library [p. 270]. Comparison of the DNA sequence of the cDNA clone and the genomic clone revealed the location of introns and exons. Isolation and sequencing of the beta-globin genes from individuals with sickle-cell anemia or beta-thalassemia revealed the nature of the mutations that caused these diseases.

Task 1, 2, 3, 4, 5, 6, 7, 8, 9

INTRODUCTION

The protein that is defective in both sickle-cell anemia and in beta-thalassemia is beta-globin which is part of the oxygen-carrying molecule hemoglobin.

In over 50% of human cancers, the protein p53 has been found to be defective.

In this module you will use the amino acid sequences of these proteins as the starting point for finding their genes. You will need to find a stretch of amino acids in each protein that is not very degenerate. This means that the three-base DNA code for each of the amino acids has the fewest possible alternatives.

Once you have designed your primer based on the protein sequence, you will add a radioactive label to it. You will then use it as a probe to detect clones in the cDNA library.

You will need to select a cDNA library to screen with your probe. The choice of cDNA libraries depends on where you think there will be the largest amount of the RNA for the gene that you are looking for.

Allowing the labeled probe to find a DNA (or RNA) that it will bind to (or base-pair with) is called hybridization. Very precise conditions are needed for hybridization to succeed. If the conditions are not correct then the labeled probe may not find the DNA that it is complementary to, it may bind to lots of DNAs that it is not complementary to, or both.

Task 1, 2, 3, 4, 5, 6, 7, 8, 9

CONCEPTUAL AND TECHNICAL OVERVIEW

Amino Acid Sequences

Click here for the one-letter amino acid code:

The protein sequence of beta-globin is:
MVHLTPEEKSAVTALWGKVNVDEVGGEALGRLLVVYPWTQRFFE
SFGDLSTPDAVMGNPKVKAHGKKVLGAFSDGLAHLDNLKGTFATLSELHCDKLHVDPENFRLLGN
VLVCVLAHHFGKEFTPPVQAAYQKVVAGVANALAHKYH

The protein sequence of p53 is: MEEPQSDPSVEPPLSQETFSDLWKLLPENNVLSPLPSQAMDDLM
LSPDDIEQWFTEDPGPDEAPRMPEAAPRVAPGPAAPTPAAPAP
APSWPLSSSVPSQKTYQGSYGFRLGFLHSGTAKSVTCTYSPALN
KMFCQLAKTCPVQLWVDSTPPPGTRVRAMAIYKQSQHMTEVV
RRCPHHERCSDSDGLAPPQHLIRVEGNLRVEYLDDRNTFRHSVVV
PYEPPEVGSDCTTIHYNYMCNSSCMGGMNRRPILTIITLEDSSGNLL
GRNSFEVRVCACPGRDRRTEEENLRKKGEPHHELPPGSTKRALPN
NTSSSPQPKKKPLDGEYFTLQIRGRERFEMFRELNEALELKDAQAG
KEPGGSRAHSSHLKSKKGQSTSRHKKLMFKTEGPDSD

Primers

Designing a primer based on a protein sequence that will bind only to one cDNA and not others requires a trade-off between two parameters. The primer must be as long as possible, but it must also be as simple as possible.

Longer primers are better because they are more likely to be specific for only one stretch of DNA sequence. The problem is that the longer the primer, the more likely it is that it will cover amino acids having many possible codons. The more codons there are for each amino acid that is used, the more likely that the primer will bind to other cDNAs.

Click here to see a table of the genetic code.

Thus, to design effective primers for cDNA cloning you need to find a stretch of amino acids in the protein sequence that has low degeneracy. This means that there are a small number of alternatives for the codons that code for these amino acids.

Click here to see the genetic code arranged according to the pattern of degeneracy.

One way to look for sequences with low degeneracy is first to search the amino acid sequence for methionine (M) or tryptophan (W) which are amino acids that have only a single codon.

Then see if there are amino acids that surround the M or W for which a high proportion have 2 or 3 codons. Try to avoid any stretch that contains one of the three amino acids (R, L or S) that have 6 codons each.

The goal is for the total degeneracy to be less than 260. The total degeneracy is calculated by multiplying the number of alternatives for each codon used to make the primer.

For example, if the primer has 3 possible codons at one position, 2 possible codons at three positions, 1 possible codon at one position and 4 possible codons at two positions the total degeneracy is: $3 \times 2 \times 2 \times 1 \times 4 \times 4 = 192$.

A good length for a primer with a total degeneracy of less than 260 is 20 nucleotides. Note that this is one nucleotide less than the codons for 7 amino acids. Making the primer 20 nucleotides and not 21 means that only the first two nucleotides of the last amino acid in the region are used. Thus, if the last amino acid has four possible codons, there is no degeneracy added to the primer, because the alternative codons all change the third base, which is not used.

When designing degenerate primers use "/" between bases that go in the same position. For example if the codon has either A or G at one position, use A/G; if there are all four nucleotides at a position, use A/T/C/G.

Thus, a 20-nucleotide primer for the amino acid sequence HIKWVIP would look like this:
CAC/TATA/C/TAAA/GTGGGTA/C/GTATA/C/TCC
The degeneracy of this primer is: $2 \times 3 \times 2 \times 1 \times 4 \times 3 \times 1 = 144$.

Write the primer sequence in the same direction (5' to 3') as the amino acid sequence. Because cDNA is double-stranded the primer will hybridize to the strand that is complementary to the strand that codes for the protein.

Task 1, 2, 3, 4, 5, 6, 7, 8, 9

Radioactive Labeling

Detecting successful binding of a probe to its complementary cDNA requires putting a label on the probe. One of the most widely used labels in molecular biology is radioactivity. This is usually in the form of a radioactive atom incorporated into a biologically active molecule.

For hybridization, a radioactive variant of phosphorus, ^{32}P, is frequently incorporated into the probe. When ^{32}P decays, it emits a beta particle which can be detected on X-ray film.

To incorporate ^{32}P into a primer for cDNA library screening, you will use a reaction catalyzed by the enzyme "kinase."

The kinase transfers a phosphate molecule containing ^{32}P from ATP to the nucleotide on one end of the primer. To be able to use ATP as a donor molecule, the ^{32}P containing phosphate has to be in the gamma position of the ATP. This is why it is referred to as "gamma-^{32}P-ATP."

Click here to view an animation that shows the radioactive labeling reaction.

9

Although the amounts and intensities of radioactive compounds used in most molecular biology procedures are quite small compared to medical or industrial uses, safety precautions must be exercised. In GDL, of course there is no real risk, but to encourage good habits, safe practices for handling radioactivity are included.

Because Plexiglass absorbs radioactive beta emissions, the ^{32}P-ATP is kept behind a Plexiglass shield, and the labeled primer is carried in a Plexiglass holder.

Any materials that have been in contact with radioactivity should be disposed of in the radioactive trash can.

Task 1, 2, 3, 4, 5, 6, 7, 8, 9

cDNA Libraries

cDNA libraries are usually made from RNA isolated from specific organs or tissues.

The choice of organ or tissue usually depends on where the RNA of interest is made. If it is known that a particular organ or tissue has a high level of the protein one is interested in, there will likely be a high level of RNA for that protein in the same place.

Because beta-globin is a constituent of hemoglobin, a cDNA library enriched in beta-globin cDNAs should be made from a tissue in which hemoglobin is made. The largest amount of hemoglobin is found in red blood cells. Unfortunately, these cells lack nuclei and therefore cannot be used as a source of RNA, despite the fact that they contain hemoglobin protein. An alternative source is the spleen, which is a good source of erythroblasts which are precursors to red blood cells.

The p53 protein is found at low levels in almost all cells. Thus, it is likely that p53 cDNA can be found in a cDNA library made from just about any tissue.

The enzyme reverse transcriptase is added to the purified RNA, along with the four nucleotides (A, T, C, G) an energy source, and the correct buffer.

Another essential ingredient is a primer that is complementary to the RNA. To make cDNA from all messenger RNAs (mRNAs) a primer of T residues, also known as oligo dT, is frequently used. All mRNAs contain a stretch of A residues (the poly-A tail) to which an oligo dT primer can base-pair.

Reverse transcriptase copies RNA into the complementary DNA (cDNA). To make double-stranded cDNA, the other strand is usually made using DNA polymerase.

The double-stranded cDNA is ligated into a vector. This can be either a plasmid or derived from a bacteriophage. You will learn about plasmid cloning vectors in the Subcloning module.

For plasmid vectors, the cDNA in the vector is transformed into bacteria and the bacteria are spread on petri dishes containing a hardened nutrient solution.

The colonies that grow up on the petri dishes comprise the cDNA library. Each colony is composed of bacteria carrying a plasmid containing a cDNA made from the reverse transcriptase reaction.

While each colony contains a plasmid with a different product from the reverse transcriptase reaction, more than one colony can contain cDNAs from the same species of RNA.

For example, if a gene makes a lot of RNA because the cell needs a lot of that protein, then that RNA species will constitute a large proportion of the RNA in that cell. In this case, one would expect to find many colonies with plasmids that contain cDNAs of this highly abundant RNA.

To screen the cDNA library one needs to transfer a bit of each of the bacterial colonies to a nylon filter that will bind the plasmid DNA.

Click here to view a movie showing how cDNA library filters are prepared.

10

The first step is to place the nylon filter on the petri dish containing the cDNA colonies. Some cells from each of the bacterial colonies stick to the filter.

The filter is marked so that the colonies that hybridize to the probe can be found after the hybridization.

To make a replicate filter, a second filter is placed on the same petri dish. It will also pick up some cells from each bacterial colony. The marks left in the petri dish from the first filter are used to mark the second filter in the same places as the first filter.

The filters are then treated to separate the two DNA strands and make them stick to the filters.

Task 1, 2, 3, 4, 5, 6, 7, 8, 9

Replicate Filters

One of the biggest problems in cDNA cloning is knowing when the probe has really bound to the gene you are looking for, or when it is just bound to something by chance. A good way to distinguish between these possibilities is to probe two identical filters.

The spots that show up on the same places in both filters are much more likely to represent real hybridization than spots that are only on one of the two filters.

The two identical filters are referred to as replicates. The two replicates have markings that allow you to line them up after the hybridization.

Task 1, 2, 3, 4, 5, 6, 7, 8, 9

Hybridization

The process of allowing one strand of a nucleic acid (DNA or RNA) to find and base-pair with its complementary strand is called hybridization.

Nucleic acids also have a tendency to bind to other nucleic acids to which they are only partially complementary. In this case some or most of the bases don't match the bases opposite them. This is called non-specific hybridization or background hybridization.

It is challenging to get hybridization to occur within a reasonable time frame and with a minimum of background hybridization. Specific hybridization requires a certain set of conditions, beginning with the use of a special solution.

This special hybridization solution generally contains large amounts of nonlabeled RNA or DNA whose role is to bind to all molecules that might nonspecifically bind nucleic acids, before the labeled probe binds to these molecules. The hybridization solution usually also contains a substance that reduces the effective volume in which the hybridization takes place.

Two variables can be easily manipulated to alter the hybridization conditions - salt concentration and temperature.

A dynamic tension exists between the hydrogen bonds that hold two complementary nucleic acid strands together and their negatively charged phosphate backbones which tend to push them apart. Raising the salt concentration of the hybridization solution lessens the repulsive force of the phosphates. Although this stabilizes the base-pairing, it also allows more mismatched base-pairing to occur.

To reduce the number of variables, in GDL the hybridization solution contains an optimal salt concentration that reduces the amount of mismatched base-pairing.

For a particular salt concentration, changing the temperature will alter the ratio of correct hybridization to mismatched base-pairing. As the temperature is raised, mismatched base-pairs will tend to pull apart, because they have weak or no hydrogen bonds holding them together. The converse is also true: As the temperature is lowered, more mismatching will occur.

Of course, there is an upper limit to the temperature that can be used. Above a certain temperature, even the bonds between correctly paired bases will be broken. The salt and temperature of a hybridization define the stringency conditions of the hybridization. Low salt and high temperature are high stringency conditions, high salt and low temperature are low stringency conditions.

Task <u>1</u>, <u>2</u>, <u>3</u>, <u>4</u>, <u>5</u>, <u>6</u>, <u>7</u>, <u>8</u>, <u>9</u>

Temperature and Time of Hybridization

In the case of the GDL hybridization solution, for primers of 20 nucleotides in length, with a degeneracy of between 128 and 260, a good hybridization temperature is 37degrees C.

For primers with a total degeneracy of less than 128 then the temperature can be as high as 42degreesC.

Beyond a certain minimum, the time of hybridization is not critical. The probe needs adequate time to find its complementary strand. This depends on several factors including abundance of the complementary strand and amount of probe.

Normally, hybridizations are performed overnight. In GDL the minimum hybridization time is 4 hours.

Task <u>1</u>, 2, <u>3</u>, <u>4</u>, 5, <u>6</u>, <u>7</u>, <u>8</u>, <u>9</u>

Prehybridization

The purpose of prehybridization is to prepare the filters for hybridization with radioactive probe. This process begins with exposing the filters to an abundance of DNA or RNA, thereby reducing the likelihood that the probe will bind to the wrong places.

Prehybridization is performed by adding <u>hybridization solution</u> to the filters and incubating them in the hybridization oven before adding the radioactively labeled probe.

To be effective, prehybridization should be performed at a temperature that is at least as high as the hybridization temperature. Prehybridization is often performed at 5 degrees warmer than the hybridization temperature.

Prehybridization is normally performed for at least 1 hour.

Task <u>1</u>, 2, <u>3</u>, <u>4</u>, <u>5</u>, <u>6</u>, <u>7</u>, <u>8</u>, <u>9</u>

Filter Washing

Washing unbound probe from filters after hybridization is an essential step to reduce background.

The first step is to pour off the hybridization solution from the roller bottle. Because the solution also contains radioactively labeled probe, it must be poured into the liquid radioactive waste container.

Effective washing of filters depends on the same parameters as hybridization: salt concentration and temperature. In GDL, the salt concentration of the wash solution is automatically adjusted for the degeneracy of the probe used.

The temperature of washing depends on the degeneracy of the primer. For primers of 20 nucleotides in length, with a degeneracy of between 128 and 260, a good washing temperature is 42degreesC.

For primers with a total degeneracy of less than 128 then the temperature can be as high as 50 degrees C.

Filters should remain in the wash solution for 30 minutes.

Task <u>1</u>, 2, <u>3</u>, <u>4</u>, 5, <u>6</u>, <u>7</u>, <u>8</u>, <u>9</u>

PROCEDURES

Task 1: Design primers to probe the cDNA library.

1. Your first task is to design the primers that you will use to probe the cDNA library. The DNA sequence of the primers must reflect the amino acid sequence of the protein that you are interested in.

2. To enter the primer sequence click on the arrow on the plexiglass shield in front of the oligonucleotide rack. On the menu that appears select, "New oligonucleotide sequence." Alternatively, you can choose one of the test sequences.

3. In the dialog box, use the four yellow keys (A, T, C or G) to enter the nucleotides for your primer in the 5′ to 3′ direction. For degenerate primers use "/" to separate nucleotides at the same position. Your primer should be 20 bases long.

4. Click DONE when the primer is entered. Primers are at a concentration of 0.5 milligram/milliliter. The sequence of the primer that you designed will be automatically entered into your notebook.

5. If you have made a mistake dispose of the primer tube by dragging it to the trash can.

Task 2: Set up the labeling reaction.

1. To attach a radioactive label to the primer use the pipettor to add 20 microliters of gamma ^{32}P-ATP to the tube that contains your primer. The ^{32}P -ATP is kept behind a Plexiglass shield to protect the user from the radioactivity.

2. Use the pipettor to add 7.5 microliters of 10x (concentrated tenfold) kinase buffer found in the reagent rack. Buffers are used at 1X concentration.

3. To select the kinase, click on the arrow on the enzyme cooler. On the enzyme selection menu that appears, select T4 polynucleotide kinase.

4. Use the pipettor to add 5 microliters of kinase to the primer tube.

Task 3: Incubate the labeling reaction.

1. Drag the primer tube to the heat block. Turn on the heat block . Set the temperature for 37 degreesC.

2. Set the time of incubation for 45 minutes by clicking on the timer.

3. While the reaction is incubating, click here to view an animation of the radioactive labeling reaction.

4. When the reaction is complete, drag the tube to the Plexiglass radioactive tube container at the right of the screen. This will shield you from the radioactivity.

Task 4: Select cDNA libraries.

1. Press NEXT to move to the next screen. You will find 4 circular cDNA library filters. Click here to find out how cDNA library filters are prepared.

2. You can select two different libraries to screen with your probe. To select a cDNA library, click on the arrow between the filters.

3. On the selection menu choose the <u>cDNA library</u> that you wish to have on the two filters on the left. Choose the library made from a tissue that is likely to contain the gene that you are looking for.

4. Repeat this process to select a library for the two filters on the right. The two filters on the left of the screen are identical to each other, as are the ones on the right. They are known as <u>replicates</u>. They are very useful in determining what is real hybridization and what is background.

Task 5: <u>Set up the prehybridization.</u>

1. To set up the <u>prehybridization</u> use the <u>forceps</u> to put each of the filters into the <u>roller bottle</u>

2. Use the <u>glass pipette</u> to transfer 10 ml of <u>hybridization solution</u> from its flask to the roller bottle with the filters in it.

3. To prehybridize the filter, cap and drag the roller bottle to the <u>hybridization oven</u>. The door will automatically open and the roller bottle will be placed in its holder.

4. Click on the buttons on the left side of the hybridization oven to set the <u>temperature</u> on the hybridization oven. For a primer with degeneracy of between 128 and 260, a good prehybridization temperature is 42degreesC. For primers with a total degeneracy of less than 128 then the temperature can be as high as 47degreesC.

5. On the timer set the time for the <u>prehybridization</u>. One hour is usually sufficient.

6. When the prehybridization is complete, click on the hybridization oven to have the roller bottle return to the bench.

Task 6: <u>Add probe to the roller bottle.</u>

1. Use the pipettor to add 25 microliters of the radioactively labeled probe to the roller bottle with the filters.

2. Drag the capped roller bottle into the hybridization oven.

3. Set the temperature on the hybridization oven. The <u>temperature</u> depends on the degeneracy of the probe that you designed. For a probe of degeneracy between 128 and 260 an appropriate temperature is 37degreesC. If the degeneracy is less than 128 then the temperature can be set as high as 42degrees C.

4. On the timer set the <u>time</u> for the hybridization. It must be for at least 4 hours.

5. When the hybridization is complete, click on the hybridization oven door to bring the roller bottle back to the bench.

Task 7: <u>Wash the filters.</u>

1. To <u>wash the filters</u>, take off the cap, then drag the roller bottle to the radioactive waste bottle and pour off the hybridization solution.

2. Use the glass pipette to transfer 10 ml from the beaker with <u>wash solution</u> to the roller bottle.

3. Put on the cap and drag the roller bottle back into the hybridization oven.

4. Set the <u>temperature</u> on the hybridization oven. For a probe of degeneracy between 128 and 260 an appropriate temperature is 42degreesC. If the degeneracy is less than 128 then the temperature can be set as high as 50degreesC.

5. On the timer set the <u>time</u> for the wash. One hour is usually sufficient.

6. When the wash is complete, click on the door to bring the roller bottle to the bench.

Task 8: <u>Place filters in the X-ray cassette</u>.

1. Click NEXT to move to the next screen.

2. Take the cap off of the roller bottle and use the forceps to drag the filters one at a time, to the <u>X-ray cassette</u>.

3. Click the top of the X-ray cassette. After a few moments, you will see the filter exposing the X-ray film.

4. Take a picture of the filter by clicking on the film icon. The picture will be entered into your notebook.

5. Close the window to return to the bench.

Task 9: <u>Interpret the results</u>.

1. To decide which spots represent real hybridization to cDNAs and which are background, rotate one of the replicate filters until the marking spots line up.

The number of spots that line up with each other will depend primarily on the probe that you designed and the cDNA library that you selected. If the probe is truly complementary to beta-globin or p53, then the number of clones that are found on both filters is proportional to the amount of RNA that codes for that protein in the tissue that was used to make the cDNA library.

For example, if a probe is complementary to globin and there are lots of positive clones in the spleen cDNA library, this indicates that globin RNA is abundant in the spleen as compared to other RNAs found in spleen.

2. Once you have found the spots located in the same positions on both filters note their locations in your notebook. You will refer to this information when you begin another module that requires the cDNA. Modules in which you can use the results from this module are: Restriction Mapping, PCR, Southern Blotting and Northern Blotting.

<u>TROUBLESHOOTING</u>

If you found no spots on that appear on both of your filters the most likely causes are:
*   You forgot to put the probe into the roller bottle.
*   You failed to perform the kinase reaction to add the radioactive label to your probe, you did not use the right enzyme, or you forgot to add one of the ingredients.
*   You forgot to put the kinase reaction in the heat block.
*   You didn't set the temperature correctly for the kinase reaction.
*   You didn't set the timer correctly for the kinase reaction.
*   You forgot to add hybridization solution to the roller bottle.
*   Your probe was the wrong sequence. It must be complementary to either globin or p53.
*   The degeneracy of the primer was too high. A degeneracy higher than 260 will result in no (or very weak) hybridization.
*   You didn't leave the roller bottle in the hybridization oven long enough.
*   The temperature of the hybridization or wash was too high (higher than 65degreesC).
*   The tissue from which the cDNA library was made doesn't make the RNA that you are interested in.

If there are fewer than expected positive spots that appear on both of your filters the most likely cause is:
*   The temperature of hybridization or wash was too high for the degeneracy of the probe.

If you found too many spots that don't line up, in other words there was too much background - the most likely causes are:
- The hybridization oven temperature was set too low for prehybridization or hybridization.
- The wash temperature was too low.
- You forgot to do the prehybridization
- You forgot to wash the filters after hybridization.
- There were some mistakes in the sequence of the probe.
- The degeneracy of the probe was too high.

EQUIPMENT AND REAGENT INFORMATION

Pipettor

Drag the automatic pipettor from its rack and a new disposable tip will appear. The tip will be automatically discarded after each use.

Click on the dark rectangular window on the pipettor to get a pop-up slider. Select the volume by adjusting the slider, by using either the sliding needle or the arrow buttons for fine control. Click OK or anywhere outside the slider panel to dismiss it.

Drag the pipettor to the rack containing tubes. When the tip of the pipettor is over a tube, the contents of the tube will be visible. This tells you which tube the pipettor tip is in.

When you release the mouse, if the volume is set within the correct limits, the reagent will be drawn up into the pipettor tip. If not, you will get an error message and will need to reset the volume.

Drag the pipettor to the tube to which you want to add the reagent. When the pipettor tip is over the tube, release the mouse. What was in the pipettor tip will now be added to that tube.

Heat Block

Turn on the heat block by clicking the ON/OFF switch. Click the control knob on the heater to set the temperature. A slider appears that will allow you to set the temperature.

If you push GO on the timer and you have forgotten to turn on the heat block you will get an error message.

Timer

When you click on the timer a slider will appear. Set the slider on the timer for the desired time. Click the GO button on the timer to start the timer. In GDL, time is speeded up, so the timed process will take only a few seconds.

If you need to adjust anything, click outside the time slider to dismiss it without starting the timer.

Forceps

To use the forceps, drag them toward the gel or filter.

Release the mouse when the tip of the forceps is over the filter. Now drag the forceps – the forceps will close on the filter and drag it.

Release the forceps when the filter is in the desired location.

Glass pipette

Drag the glass pipette from the bench.

Click on the volume element at the top of the glass pipette to get a pop-up slider.

Select the volume by adjusting the slide, by using either the sliding needle or the arrow buttons for fine control. Click OK or anywhere outside the slider panel to dismiss it.

Drag the glass pipette to a bottle or flask. When you release the mouse, if the volume is set within the correct limits, the reagent will be drawn up into the glass pipette. If not, you will get an error message and will need to reset the volume.

Drag the glass pipette to the tube or roller bottle to which you want to add the reagent. When the glass pipette is over the tube or roller bottle, release the mouse. The liquid in the glass pipette will now be added to the tube or roller bottle.

Hybridization solution

The hybridization solution for degenerate primers has the following composition: 6x SSC, 1x Denhardt's, 100 ug/ml tRNA and 0.05% pyrophosphate.

A stock solution of 20x SSC is made by dissolving 175.3 grams of sodium chloride and 88.2 grams of sodium citrate in 800 ml of water. The pH is adjusted to 7.0 with the addition of sodium hydroxide, then the volume is adjusted to 1 liter.

A stock solution of 50x Denhardt's reagent is made by dissolving 5 grams of Ficoll, 5 grams of polyvinylpyrrolidone, and 5 grams of bovine serum albumin in 500 ml of water. Denhardt's reagent is used to reduce the effective aqueous volume of the hybridization solution.

Hybridization wash solution

The hybridization wash solution for degenerate primers has the following composition: 6x SSC, 0.05% pyrophosphate.

Roller bottle

Roller bottles are used because they allow hybridization to take place in a small volume while permitting the probe to contact the filter evenly over its entire surface.

To cap the roller bottle, drag the cap from the bench and release it over the roller bottle, it will automatically drop onto the roller bottle.

To pour off hybridization solution, drag the roller bottle to the liquid radioactive waste beaker. When the roller bottle is released, it will tip up and pour off the liquid.

Hybridization oven

Click on the hybridization oven door to open or close it. Set the oven temperature by clicking on the buttons to the left of the door.

Turn on the oven by clicking the ON/OFF switch. Set the timer. The timer will not run if the oven is not turned on.

X-ray cassette

The forceps must be used to transfer the filter from the roller bottle to the X-ray cassette.

To activate the X-ray cassette, click on the top part of the cassette.

Restriction Mapping

OBJECTIVES

The procedures in this lab will enable you to:

- **Cut DNA with restriction enzymes.**
- **Separate DNA fragments by gel electrophoresis.**
- **Make a restriction map.**

PURPOSE

Restriction mapping is used to determine where restriction enzymes cut a piece of DNA.

You can generate a restriction map of the cDNA you isolated in the cDNA Cloning module or of a PCR fragment you amplified in the PCR module. You can use the restriction map to decide which fragments to subclone for sequencing or for insertion into a protein expression vector.

To get started immediately, go to Task 1.

TABLE OF CONTENTS

Background
Introduction
Conceptual and technical overview
Task 1: Select the DNA to be cut with restriction enzymes.
Task 2: Add restriction enzymes to the DNA.
Task 3: Incubate the restriction digest.
Task 4: Add running dye to the digested DNA.
Task 5: Load the agarose gel.
Task 6: Run the gel.
Task 7: Determine the size of the restriction fragments.
Task 8: Make a restriction map.
Troubleshooting
Equipment and reagent information

BACKGROUND

A "restriction map" shows the places where different restriction enzymes cut a piece of DNA. Restriction enzymes are a way for bacteria to defend themselves. A virus that is invading the bacterial cell will have its DNA cut up by the bacterial restriction enzyme. The bacterium protects its own DNA from being cut up by methylating it in special places, thereby preventing the restriction enzyme from being able to cut the DNA.

Three types of bacterial protection systems use restriction enzymes. The type II system has separate proteins that act as restriction enzymes and as methylases. The other two kinds, type I and type III, use the same protein to methylate and cleave the DNA. These systems don't cut the DNA at the restriction site where they bind. They also require ATP to cut the DNA. For these two reasons they are not very useful for DNA manipulation.

Restriction enzymes recognize specific sequences of base-pairs in the DNA. For example the restriction enzyme *Eco*RI (the Eco stands for "*E. coli*," which is the bacterium that makes it) recognizes the sequence:

5´...GAATTC...3´
3´...CTTAAG...5´

This sequence is a palindrome, which means that the sequence is the same on both strands when read in the 5′ to 3′ direction . When *Eco*RI cuts at this restriction site it produces staggered ends.

```
5´...G          AATTC...3´
3´...CTTAA          G...5´
```

These staggered ends are the same no matter where the DNA comes from.

APPLICATIONS

Because the staggered ends generated by restriction enzymes will anneal to each other through base-pairing, they are one of the key tools in making recombinant DNA. There are many applications of recombinant DNA technology, from protein production to gene therapy.

Restriction enzymes combined with Southern blotting are used extensively in genetic analysis. In particular, DNA markers known as restriction fragment length polymorphisms (RFLPs) are generated by restriction enzymes. A RFLP occurs between two individuals when their DNA is cut with the same restriction enzyme and hybridized with the same probe and the resulting bands are at different positions. The RFLP exists because random mutations, not necessarily within genes, alter the cleavage sites of restriction enzymes.

RFLPs can be used as genetic markers to map disease genes. This was how the gene for Huntington disease was first mapped. RFLP analysis is also used in genetic testing for diseases.

Task 1, 2, 3, 4, 5, 6, 7, 8

INTRODUCTION

The cDNA that you isolated in the cDNA module was made in the pUC19 cloning vector. To map the cDNA you need to cut it with different restriction enzymes. You will then determine how far the resulting restriction fragments migrate after separating them using agarose gel electrophoresis. By comparing these data to the distance migrated by restriction fragments of known length (called the gel marker), you will be able to determine the length of the cDNA restriction fragments. Knowing the length of the fragments will enable you to determine where that restriction enzyme cuts the DNA. By doing this with several enzymes, either alone or together, you will be able to generate a restriction map.

Finding where restriction enzymes cut in a piece of DNA is useful for sequencing. For nonautomated sequencing, pieces of DNA need to be less than 500 base-pairs (bp) in length. This is because with manual sequencing one is generally only able to read 250-300 bases. If you sequence a 500 bp piece of DNA in both directions, then you can find where the two sequences overlap. Thus, a first step in manual sequencing is to subclone fragments of around 500 bp. To identify suitable fragments for subcloning requires restriction mapping.

Cutting PCR amplified DNA fragments with restriction enzymes can determine if a restriction site is present in a mutant form of the DNA that is absent in the normal DNA (or vice versa).

Task 1, 2, 3, 4, 5, 6, 7, 8

CONCEPTUAL AND TECHNICAL OVERVIEW

Features of Cloning Vectors

Plasmid cloning vectors contain several features:

- An origin of replication, which allows the plasmid to be replicated inside a bacterial cell.

- At least one selectable marker that confers resistance to an antibiotic so that you can select the bacteria that has taken up the plasmid. Any bacteria that does not contain the vector will die in the presence of antibiotic.

- One or more restriction sites which occur only once on the plasmid. These allow you to insert foreign DNA into the plasmid without cutting something important (like the origin of replication).

- A means of identifying which plasmids have incorporated foreign DNA; in other words a means to tell recombinant plasmids from nonrecombinant plasmids.

Task <u>1</u>, <u>2</u>, <u>3</u>, <u>4</u>, <u>5</u>, <u>6</u>, <u>7</u>, 8

pUC19 Plasmid

The cDNA library used in GDL was made using the plasmid vector pUC19. Each of the cDNAs was inserted into the SmaI site of the pUC19 polylinker.

In the pUC19 plasmid, the origin of replication is marked "ori." <u>Click here to view a detailed map of the pUC19 plasmid.</u>

The selectable marker is a gene that confers resistance to the antibiotic ampicillin. This marker is on the part of the plasmid labeled "amp^r".

There are 13 unique restriction sites placed close together in a region known as the polylinker. In addition, there are several other unique restriction enzyme sites at other locations on the pUC19 plasmid.

Recombinant plasmids are identified by the blue/white selection scheme which involves interrupting the lacZ gene. This process is described in detail in the Subcloning module.

Task <u>1</u>, <u>2</u>, <u>3</u>, <u>4</u>, <u>5</u>, <u>6</u>, <u>7</u>, 8

Restriction Enzymes: Amount to Use

The restriction enzymes should be added so that they make up no more than 1/10 of the final volume. This is because they are stored in glycerol, and adding too much glycerol to the DNA tube will inhibit the reaction.

For the 20 microliter reactions used in GDL, 2 microliters of restriction enzyme is appropriate.

You generally need to add at least 1 microliter of a restriction enzyme to get it to cut efficiently.

In GDL, if you want to add more than two restriction enzymes to the same tube, you can use 1 microliter of each.

Task <u>1</u>, 2, 3, 4, 5, 6, <u>7</u>, 8

Temperature of Restriction Digests

Many restriction enzymes come from bacteria that live inside mammals. Thus they work best at 37 degrees Celsius.

At room temperature (22 degrees Celsius) most restriction enzymes don't cut DNA very well.

Task <u>1</u>, 2, <u>3</u>, <u>4</u>, 5, 6, <u>7</u>, 8

Partial Digests

If restriction digests are performed at lower temperatures than they should be or for shorter time periods than they should, the enzyme will not be able to cut at every restriction site. The result is a partial digest.

In a partial digest, because the restriction enzyme does not cut at every site, it is very difficult to interpret the digestion pattern.

In general, leaving a restriction digest for longer than the recommended time has no negative effect.

Task 1, <u>2</u>, 3, <u>4</u>, <u>5</u>, <u>6</u>, <u>7</u>, 8

Agarose Gel Electrophoresis

Gel electrophoresis is a means of separating DNA fragments and allows you to calculate their approximate sizes.

The size of a DNA fragment is a very useful means of identifying it when trying to sort out mixtures of DNA fragments.

DNA has a negative electrical charge provided by the phosphate groups on the DNA backbone. When DNA is in an aqueous solution, hydrogen molecules are removed from the phosphates resulting in a net negative charge.

Because there is one phosphate for each nucleotide, the negative charge is directly proportional to the length of the DNA fragment.

Agarose gel electrophoresis is performed in a gel box into which the gel is placed. A current is run through a buffer solution surrounding the gel, with the negative electrode near where the DNA is added to the gel, and the positive electrode at the other end of the gel box.

DNA is attracted to the positively charged electrode, with smaller pieces of DNA moving more easily and quickly through the pores of the gel than larger pieces of DNA.

Agarose gels are typically used to separate restriction fragments that are larger than 200 base pairs (bp).

A gel of the size used in this module normally runs at approximately 100 Volts (V), which produces about 120 milliAmperes (mA) of current.

Task 1, 2, 3, 4, 5, 6, 7, 8

Ethidium Bromide

Ethidium bromide binds to DNA by intercalating between the bases.

If ethidium bromide is added to an agarose gel it will bind to the DNA fragments as they migrate through the gel.

When illuminated by UV light, ethidium bromide glows orange, making it easy to visualize the location of the DNA fragments in the gel.

Caution: ethidium bromide is a potent mutagen and should be handled with extreme care .

Task 1, 2, 3, 4, 5, 6, 7, 8

PROCEDURES

Task 1: Select the DNA to be cut with restriction enzymes.

1. Click the arrow on the tube rack to get the DNA selection menu.

2. Select a DNA by highlighting a menu option and releasing the mouse.

3. When you roll over the DNA tube its contents will be listed. Each DNA tube contains 1microgram of the selected DNA at a concentration of 1 microgram/microliter as well as 2 microliters of 10x buffer and 15 microliters of distilled water.

4. To get another tube, click the arrow again and the menu will reappear. You can select as many as 10 DNA tubes.

5. To determine the restriction map of a cDNA you will want to compare its restriction fragments with those from the pUC19 plasmid vector alone. For each restriction enzyme you choose to use, you should select pUC19 plasmid for one of your DNA tubes.

6. If you make a mistake, drag the tube to the trash can underneath the bench.

Task 2: Add restriction enzymes to the DNA.

1. To select restriction enzymes, click the arrow on the enzyme cooler. When you make your selection from the menu, a tube will appear in the cooler, containing the selected enzyme.

2. Use the pipettor to transfer restriction enzymes to the DNA tube.

3. Adjust the volume on the pipettor for the correct amount of restriction enzyme. Two microliters is an appropriate amount for a 20 microliter reaction.

4. You can add more than one restriction enzyme to each tube. The final contents of each tube will be automatically recorded in your notebook.

Task 3: Incubate the restriction digest.

1. After adding the restriction enzymes, drag the DNA tubes one by one to the heat block.

2. When all tubes are in the heat block, turn it on by clicking the ON/OFF switch.

3. Set the temperature for 37 degrees Celsius.

4. Set the timer. Most restriction enzymes will completely digest DNA within 1 hour. If less time is allowed for the digestion, then you will get a partial digestion.

5. While the reaction digest is incubating, click here to access an animation that will show you what's going on in the tubes.

Task 4: Add running dye to the digested DNA.

1. When the timer signals the end of the incubation, press the NEXT button to move down the bench to where the next rack is located.

2. Drag the tubes, one by one, to the empty rack.

3. To prepare your digested DNA for running on a gel you must add running dye. The running dye or "blue dye" is in the small rack.

4. Use the pipettor to add blue dye to each of your digested DNAs. For a 20 microliter DNA reaction, add 2 microliters of blue dye.

Task 5: Load the agarose gel.

1. You will now use agarose gel electrophoresis to separate the DNA fragments. Click here to view a movie on how to prepare an agarose gel.

2. Use the pipettor to transfer 10 microliters of marker DNA found in the small rack to the first gel well on the left. When the pipettor tip is over the well, release its contents by releasing the mouse button.

3. Use the pipettor to transfer 10 microliters of the digested DNA (to which you have added blue dye) to a different well of the gel. The contents of each well are revealed when the pipettor rolls over them and they will be automatically entered into your notebook.

Task 6: <u>Run the gel</u>.

1. When all of the wells are loaded, click the NEXT button.

2. Drag the cover onto the gel box. It is correctly positioned when you hear it click into place.

3. Adjust the <u>voltage</u> by clicking on the voltage/amperage indicator. 100 volts is an appropriate voltage for this gel.

4. Turn on the power supply by clicking the ON/OFF switch. A new window will open to give you a good view of the gel inside the gel box.

5. After a few moments, an ultraviolet light will illuminate the gel in which <u>ethidium bromide</u> binds to the DNA fragments and glows orange.

6. The voltage can be adjusted while the gel is running.

7. You can stop and resume the electrophoresis at any time. The goal is to get maximum separation of the fragments without letting any of them run off the bottom of the gel. Fragments that do run off the end of the gel are lost.

8. Click on the camera icon to take a picture of the gel. The picture will be entered into your <u>notebook</u>. You can take up to three pictures of each gel.

9. If you would like to digest another set of DNAs with the same or other enzymes, close the window to return to the bench.

Task 7: <u>Determine the size of the restriction fragments</u>.

1. From your notebook find the pictures you took of the gel. Open the one that shows the best separation of the restriction fragments. By determining the sizes of fragments generated by each of the restriction digests, you will be able to make a restriction map. Subsequent digests based on this first map will allow you to refine and improve the map.

2. To get an estimate of fragment size, compare the position on the gel of a DNA fragment with the position of the <u>gel marker</u> DNA bands. Those that appear at about the same place have about the same size.

3. To get a more accurate size of a restriction fragment, use the ruler to measure the distance from the starting well to each of the <u>gel marker</u> fragments and plot this information on a piece of graph paper or in a graphing program.

4. Make the Y axis the logarithm of fragment length and the X axis the distance migrated. Draw a straight line through the plotted points.

5. Next, use the ruler to determine the distance each fragment from your digested DNA has migrated in the gel.

6. Plot each of these distances on the graph you made of the gel marker DNA size versus distance and determine the approximate size of the fragment and make a table of the sizes of the fragments generated by each enzyme.

Task 8: <u>Make a restriction map</u>.

1. For each restriction enzyme that you used, compare the number of fragments obtained for the cDNA (inserted into pUC19) with the number of fragments obtained for <u>pUC19</u> alone.

2. If the number of fragments is the same it means that this enzyme does not cut within the cDNA. If the number of fragments is greater for the cDNA than for pUC19, it means there is at least one restriction site for that enzyme within the cDNA.

Example A:

You cut a sample cDNA with *Bam* HI and get two fragments, one of approximately 3200 bp, the other 300 bp. First, this tells you that the combined length of your cDNA and pUC19 is 3200+300=3500 bp. Because you know the length of pUC 19 is about 2,700 bp then you can estimate the size of the cDNA as 3500-2700=800 bp. You also know that there is a unique Bam HI site in pUC19 right next to the SmaI site where the cDNA is inserted. Thus the smaller fragment must be from the BamHI site in the polylinker to a BamHI site that is 300 bp from that side of the cDNA. Therefore, the longer fragment is the remaining 500 bp of the cDNA still attached to the 2700 bp of pUC 19.

To make a map use your own drawing program to make a circle for pUC 19, with a cDNA inserted into it of about 800 bp. Mark the location of the Bam HI sites on the polylinker and on the cDNA. This is a straightforward example in which there are no ambiguities.

Example B:

You cut your cDNA with Eco RI and get 3 fragments: 2,900 bp, 400 bp and 200 bp. You know that Eco RI cuts the polylinker near the Sma I site, and on the opposite side of the cDNA from the Bam HI site in the polylinker. Therefore, there must be two Eco RI sites in the cDNA. What you can't tell is whether one of the sites is 200 bp from the Eco RI site in the polylinker and the other one is 400 bp from that site, or one of the sites is 400 bp from the Eco RI site in the polylinker and the other site is 200 bp from that site.

To resolve this sort of ambiguity you need to do another experiment. You cut the DNA with both Bam HI and Eco RI. Because you know that Bam HI cuts 300 bp from the 5´ side of the polylinker, in a double digest either the 200 bp fragment generated by the Eco RI digest will be cut in two by Bam HI, or the 400 bp fragment from the Eco RI digest will be cut into a 300 bp fragment and a 100 bp fragment by Bam HI. With this information you will be able to determine the order of the two Eco RI sites relative to the Eco RI site in the polylinker.

3. With the data you generated in this module make a restriction map, using the examples as a guide. You will probably have to do additional experiments to make a complete map.

4. To determine the sequence of this cDNA using the Sequencing module, locate the restriction sites that will let you subclone fragments of appropriate sizes. For manual sequencing you will need fragments that are no longer than 500 bp.

5. In the Subcloning module you should subclone several pieces that overlap, to ensure that you can determine the entire sequence of the cDNA.

TROUBLESHOOTING:

If the DNA does not appear to be cut by the restriction enzyme(s) the most likely causes are:
- You forgot to put restriction enzyme into the DNA tube.
- You forgot to put the DNA tube with restriction enzyme in the heat block.
- You didn't set the temperature correctly for the restriction digest.
- You didn't set the timer correctly for the restriction digest.

If the number of bands does not make sense the most likely cause is:
- The temperature of the restriction digest was too low, resulting in a <u>partial digest</u>.
- The duration of the restriction digest was too short, resulting in a partial digest.

If smaller fragments are missing the most likely cause is:
- You didn't stop the gel in time and the small fragments ran off the gel.

EQUIPMENT AND REAGENT INFORMATION

Pipettor

Drag the automatic pipettor from its rack. A new disposable tip appears automatically. The tip will be automatically discarded after each use.

Click on the dark rectangular window on the pipettor to get a pop-up slider. Select the volume by adjusting the slider using the sliding needle or by using the arrow buttons for fine control. Click OK or anywhere outside the slider panel to dismiss it.

Drag the pipettor to the rack containing tubes. When the tip of the pipettor is over a tube, the contents of the tube will be visible. This tells you which tube the pipettor tip is in.

When you release the mouse, if the volume is set within the correct limits, the reagent will be drawn up into the pipettor tip. If not, you will get an error message and will need to reset the volume.

Drag the pipettor to the tube to which you want to add the reagent. When the pipettor tip is over the tube, release the mouse button. What was in the pipettor tip will now be added to that tube.

If you draw up something you don't want to use, drag the pipettor to the trash. Release the pipettor over the trash and the current tip will be discarded, allowing you to start over with a new tip.

Enzyme cooler

The enzyme cooler is kept in a freezer at –20 degrees Celsius and maintains this temperature for some time when placed on the bench.

Keeping enzymes cold is important because storage at higher temperatures will cause them to lose their ability to cut DNA.

Heat Block

Turn on the heat block by clicking the ON/OFF switch.

Click the control knob on the heater to set the temperature. A slider appears that will allow you to set the temperature.

Timer

When you click on the timer a slider appears. Set the slider on the timer for the desired time.

In GDL, time is speeded up, so the timed process will only take a few seconds.

Click the GO button on the timer to start the timer.

If you have forgotten to turn on the piece of equipment being used, a message will appear to remind you.

If you need to adjust anything, click outside the time slider to dismiss it without starting the timer.

Gel Marker

The Gel Marker is DNA from bacteriophage lambda that has been cut with the restriction enzyme HindIII.

This marker has been used for many years because the resulting fragments produce a nice "ladder" of known sizes.

The sizes of the marker DNA from largest to smallest are: 23,130 basepairs (bp), 9,416 bp, 6,557 bp, 4,361 bp, 2,322 bp, 2,027 bp, 564 bp, and 125 bp.

Running dye

The running dye or "blue dye" contains dextran sulfate, a substance that makes the DNA sink to the bottom of the wells in the gel.

If you didn't add it, the DNA would come out of the well and disappear in the buffer.

Blue dye also allows us to see approximately how fast the DNA is moving through the gel. Because in GDL you can see the DNA directly, you don't need to use running dye for this purpose.

Notebook

This is where you record your observations.

To review any of the gel photographs you've taken, click the gel number (such as "A02") in the records in the notebook to make the viewing window appear.

You can also use the Back and Next buttons in the viewing window to navigate through your photograph collection.

Southern Blot

OBJECTIVES

The procedures in this lab will enable you to:
- **Choose genomic DNAs for the Southern blot.**
- **Select a fragment of DNA to use as a probe.**
- **Hybridize the Southern blot with the probe.**
- **Interpret the hybridization pattern.**

PURPOSE

Southern blotting is used to compare the sizes of DNA fragments that hybridize to specific probes.

You can use Southern blotting to compare the structure of DNAs from normal individuals with that of patients with the diseases under investigation.

To get started immediately, go to Task 1.

TABLE OF CONTENTS

Background
Introduction
Conceptual and technical overview
Task 1: Choose genomic DNAs to fill lanes in gel.
Task 2: Cross-link the DNA to the filter.
Task 3: Set up the prehybridization.
Task 4: Select a radioactively labeled probe.
Task 5: Denature the probe.
Task 6: Add probe to the roller bottle.
Task 7: Wash the filter.
Task 8: Place the filter in the X-ray cassette.
Task 9: Interpret the results.
Troubleshooting
Equipment and reagent information

BACKGROUND

Most species have so much DNA that when it is cut with a restriction enzyme and separated by gel electrophoresis the individual bands cannot be distinguished from each other. Instead, they appear as a smear in a gel lane. To determine which one of the restriction fragments contains a gene of interest, a process known as Southern blotting can be used. The technique is named after its developer, E. M. Southern. The DNA is transferred by capillary action from the gel to a nylon or nitrocellulose filter. Then a probe labeled with radioactivity can be used to identify the band that is homologous to it.

The process of a probe finding the piece of DNA to which it is homologous is called hybridization. For hybridization to occur, both the probe and the homologous DNA have to be single-stranded. When a probe is made from double-stranded DNA, the bonds that hold the two strands together have to be broken, a process known as denaturing the DNA. When the probe finds its homologous DNA, the two strands will zip up to form double-stranded DNA.

APPLICATIONS

Southern blotting can be used to identify certain types of mutations in disease genes. Some point mutations will result in changes in restriction enzyme cleavage sites. Thus if genomic DNA from a patient is cut with that restriction enzyme the hybridizing band will have an altered position on a Southern blot when compared with the band from a normal person.

An example is in the beta-globin gene in sickle-cell anemia patients in which the mutation eliminates a DdeI restriction site. However, certain point mutations severely affect a gene's function, but they do not alter any restriction enzyme cleavage sites. Southern blotting cannot be used to identify these types of mutations. Other types of mutations that can be detected by Southern blotting include chromosomal deletions, insertions, translocations and trinucleotide repeat expansions.

When genomic DNA on a Southern blot is probed with fragments of a cDNA clone, the effects of mRNA processing can be detected. One of the major steps in mRNA processing is the removal of introns by splicing. Genomic DNA represents the unspliced form while the cDNA is made from the spliced mRNA. Thus, genomic fragments on Southern blots are frequently longer than the corresponding portions of cDNA because of the presence of introns in them.

Southern blotting is used extensively in genetic analysis. In particular, DNA markers known as restriction fragment length polymorphisms (RFLPs) are tracked by Southern blotting. An RFLP occurs between two individuals when their DNA is cut with the same restriction enzyme and hybridized with the same probe and the resulting bands are at different positions. The RFLP exists because random mutations, not necessarily within genes, alter the cleavage sites of restriction enzymes.

RFLPs can be used as genetic markers to map disease genes. This was how the gene for Huntington's disease was first mapped. RFLP analysis is also used in genetic testing for diseases.

Southern blotting is also used in the process known as "DNA fingerprinting" which is an important tool in forensics. With DNA fingerprinting, small amounts of DNA from blood or other tissue found at a crime scene can be used to compare with the DNA of suspects. When the DNA fingerprint is the same, there is a high likelihood that the individual was at the scene of the crime. This technique has also been used to free innocent people, falsely accused of a crime. Another use for DNA fingerprinting is to determine the relationship between a child and possible parents

Task 1, 2, 3, 4, 5, 6, 7, 8, 9

INTRODUCTION

To make a Southern blot, DNA is first cut with restriction enzymes. The resulting mixture of restriction fragments is then separated by agarose gel electrophoresis. The procedure is the same as in the Restriction Mapping module, except that the starting material is usually genomic DNA rather than plasmid DNA.

When the DNA fragments are sufficiently separated, the electrophoresis is stopped and the gel is placed on an apparatus that uses capillary action to transfer the DNA from the gel to a nylon filter. Click here to view a movie that shows how this blotting apparatus is set up.

After the transfer is complete, the gel and nylon filter are placed on the bench and the position of the gel wells are marked on the filter. In this module you will find the gel and nylon filter already placed on the bench with the wells marked. You first fill each lane of the filter with genomic DNA from different sources.

Task 1, 2, 3, 4, 5, 6, 7, 8, 9

CONCEPTUAL AND TECHNICAL OVERVIEW

Genomic DNA

Genomic DNA is the DNA found in the chromosomes of every cell of the body. Click here to view a movie showing how genomic DNA is isolated.

To make a Southern blot, the genomic DNA is first cut with restriction enzymes. The mixture of restriction fragments is then separated by agarose gel electrophoresis. Genomic DNA is very long and therefore has thousands of sites along its length for each restriction enzyme. Because there are so many different restriction fragments when genomic DNA is digested, you cannot see individual bands. Instead the DNA fragments form a smear down the gel.

The DNA fragments are then transferred from the gel to a filter. The filter is made of material that can bind DNA, such as nitrocellulose paper or nylon (nylon is used in GDL).

The Southern blotting buffer is drawn upwards by capillary action into the absorbent paper towels covering the filter. DNA molecules in the gel are also carried upwards, and they stick to the filter in the same positions that they occupied in the gel.

Task 1, 2, 3, 4, 5, 6, 7, 8, 9

Probe

Detection of a specific DNA species among the DNA fragments on the Southern blot requires a probe, which is normally a fragment of DNA that contains sequence from the gene that you are interested in.

In GDL, you can make probes from DNAs that you have generated in other modules, or you can use one of the premade probes.

If you select a DNA from another module, you need to make sure that the DNA that you choose actually contains a part of the gene that you are interested in.

If you choose a colony containing a cDNA clone from the cDNA Cloning module to use as a probe, be sure that there is a positive colony in the same place on both of the replica filters. If not, you may be selecting a background colony that will lack a cDNA from one of the genes under investigation.

If you choose a restriction fragment from the Restriction Mapping module to use as a probe, be sure that the fragment contains a part of the gene under investigation. Some of the fragments will contain only vector sequence, and these will not hybridize to any DNA fragments on the Southern blot.

Before you can use a fragment of DNA as a probe, the two strands need to be separated. This is necessary because only single-stranded DNA can hybridize effectively to the DNA on the Southern blot.

Separating the DNA strands involves breaking the hydrogen bonds that hold the bases together a process known as denaturation. Denaturation can be achieved either chemically or by heating the DNA. The temperature at which the DNA strands separate is referred to as its "melting point."

Probe denaturation is frequently achieved by heating the DNA to about 100 degrees Celsius for a few minutes.

Task 1, 2, 3, 4, 5, 6, 7, 8, 9

Radioactive labeling

To detect successful binding of the probe to a piece of genomic DNA that is complementary to it requires putting a label on the probe. One of the most widely used labels in molecular biology is radioactivity. This is usually in the form of a radioactive atom incorporated into a biologically active molecule.

For hybridization, a radioactive variant of phosphorous, ^{32}P, is frequently incorporated into the probe. When ^{32}P decays it emits a beta particle that can be detected on X-ray film.

To incorporate ^{32}P into a DNA fragment for Southern blot hybridization, a commonly used method is nick translation.

Nick translation involves the introduction of a break, or a "nick," into one strand of the DNA molecule. Next, DNA polymerase is added to begin priming at the site of the nick and to elongate the DNA molecule, adding nucleotides in the process.

One of the four nucleotides, frequently dCTP, has a ^{32}P molecule incorporated into it. Whenever dCTP is added to the growing DNA strand its radioactive label will be included in the DNA.

As the polymerase moves down the complementary strand it removes the old DNA that was there prior to the introduction of the nick. Conceptually then, the break in the DNA is moved down the DNA with the polymerase. The nick is translated (or moved) down the DNA. <u>Click here to view an animation of nick translation</u>..

Although the amounts and intensities of radioactive compounds used in most molecular biology procedures are quite small compared to medical or industrial procedures, safety precautions must be exercised. In GDL, of course there is no real risk, but to encourage good habits, safe practices for handling radioactive materials are included.

Because Plexiglass absorbs radioactive beta emissions, the labeled probe is kept behind a Plexiglass shield. Any materials that have been in contact with radioactivity should be disposed of in the radioactive trash container.

Task <u>1</u>, <u>2</u>, <u>3</u>, <u>4</u>, <u>5</u>, <u>6</u>, <u>7</u>, <u>8</u>, <u>9</u>

UV Cross-Linking of DNA to Nylon Filters

Permanent attachment of DNA to the nylon filter requires exposing it to ultraviolet (UV) light.

At the intensities of UV light normally used, very short (~ 10 second) exposure to the light is sufficient to cause the DNA to become attached or cross-linked to the nylon filter.

During gel electrophoresis, the DNA is bound by ethidium bromide. When exposed to UV light, ethidium bromide emits an orange color, making RNA visible on the nylon filter.

Task <u>1</u>, <u>2</u>, <u>3</u>, <u>4</u>, <u>5</u>, <u>6</u>, <u>7</u>, <u>8</u>, <u>9</u>

Hybridization

The process of allowing one strand of a nucleic acid (DNA or RNA) to find and base-pair with its complementary strand is called hybridization.

Nucleic acids also have a tendency to bind to other nucleic acids to which they are only partially complementary. In this case some or most of the bases don't match the bases opposite them. This is called non-specific hybridization or background hybridization.

It is challenging to get hybridization to occur within a reasonable time frame and with a minimum of background hybridization. Specific hybridization requires a certain set of conditions, beginning with the use of a special solution.

This special <u>hybridization solution</u> generally contains large amounts of nonlabeled RNA or DNA whose role is to bind to all molecules that might nonspecifically bind nucleic acids, before the labeled probe binds to these molecules. The hybridization solution usually also contains a substance that reduces the effective volume in which the hybridization takes place.

Two variables can be easily manipulated to alter the hybridization conditions - salt concentration and temperature.

A dynamic tension exists between the hydrogen bonds that hold two complementary nucleic acid strands together and their negatively charged phosphate backbones which tend to push them apart. Raising the salt concentration of the hybridization solution lessens the repulsive force of the phosphates. Although this stabilizes the base-pairing, it also allows more mismatched base-pairing to occur.

To reduce the number of variables, in GDL the hybridization solution contains an optimal salt concentration that reduces the amount of mismatched base-pairing.

For a particular salt concentration, changing the temperature will alter the ratio of correct hybridization to mismatched base-pairing. As the temperature is raised, mismatched base-pairs will tend to pull apart, because they have weak or no hydrogen bonds holding them together. The converse is also true: As the temperature is lowered, more mismatching will occur.

Of course, there is an upper limit to the temperature that can be used. Above a certain temperature, even the bonds between correctly paired bases will be broken. The salt and temperature of a hybridization define the stringency conditions of the hybridization. Low salt and high temperature are high stringency conditions, high salt and low temperature are low stringency conditions.

Task 1, 2, 3, 4, 5, 6, 7, 8, 9

Temperature and Time of Hybridization

For the hybridization solution used in GDL, a good hybridization temperature for Southern blots is 42degreesC. If hybridization is carried out at lower temperatures, there will be more background.

The time period of hybridization is not critical beyond a minimum length. Adequate time is needed for the probe to find its complementary strand. This depends on several factors, including the abundance of the complementary strand, and the amount of probe.

Hybridizations are usually performed overnight. In GDL the minimum hybridization time is 4 hours.

Task 1, 2, 3, 4, 5, 6, 7, 8, 9

Prehybridization

The purpose of prehybridization is to prepare the filters for hybridization with radioactive probe. This process begins with exposing the filters to an abundance of DNA or RNA, thereby reducing the likelihood that the probe will bind to the wrong places.

Prehybridization is performed by adding hybridization solution to the filters and incubating them in the hybridization oven before adding the radioactively labeled probe.

To be effective, prehybridization should be performed at a temperature that is at least as high as the hybridization temperature. Prehybridization is often performed at 5 degrees warmer than the hybridization temperature.

Prehybridization is normally performed for at least 1 hour.

Task 1, 2, 3, 4, 5, 6, 7, 8, 9

Filter Washing

Washing unbound probe from filters after hybridization is an essential step to reduce background.

The first step is to pour off the hybridization solution from the roller bottle. Because the solution also contains radioactively labeled probe, it must be poured into the liquid radioactive waste container.

Effective washing of filters depends on the same variables as hybridization: salt concentration and temperature. In GDL, the salt concentration of the wash solution has already been optimized.

The temperature of washing depends on the desired stringency. For a high-stringency wash, the temperature should be 60degreesC.

Filters should remain in the wash solution for 30 minutes.

If the wash temperature is above 70degreesC the probe will become unbound to the DNA and you will not see anything on the X-ray film.

If the wash temperature is below 32degreesC, the background will be excessively strong.

Task 1, 2, 3, 4, 5, 6, 7, 8, 9

PROCEDURES

Task 1: Choose genomic DNAs to fill the lanes in the gel.

1. On the bench you will find a gel on top of a filter. This is what the gel looks like after the process of transferring the DNA to filter is completed. Click here to view a movie on the DNA transfer process.

2. To choose the genomic DNAs that you want to place in each lane of the Southern blot click the arrowhead next to the gel to get the Genomic DNA selection menu. The first lane of the gel (the one at the far left) contains gel marker DNA.

3. Continue to select genomic DNAs until you have chosen all that are relevant to the disease that you are investigating, or until you have filled all the lanes. The contents of each lane will be automatically entered into your notebook.

Task 2: Cross-link the DNA to the filter.

1. Use the forceps to peel away the gel by dragging it from the raised corner. Drag it to the trash can.

2. To fix the DNA to the nylon filter you must place it under ultraviolet (UV) light. Open the UV cross-linking oven by clicking on the door.

3. Use the forceps to drag the filter to the oven.

4. Close the oven door by clicking on it.

5. Click on the oven keypad and use the slider to set the time of UV exposure for 10 seconds.

6. Turn on the oven by clicking the ON/OFF switch. A new window will open to give you a good view of the filter inside the oven.

7. Click on the film icon at the bottom of the window to take a picture of the DNA bound to the filter. The picture can be accessed from your notebook.

8. Close the window to return to the bench.

Task 3: Set up the prehybridization.

1. Use the forceps to drag the filter from the UV cross-linking oven and into the roller bottle. Place the top on the roller bottle.

2. Click the NEXT button twice to move to the screen with the hybridization oven.

3. Use the glass pipette to transfer 10 ml of hybridization solution from its flask to the roller bottle. Cap the roller bottle.

4. To prehybridize the filter, drag the roller bottle to the hybridization oven. The door will automatically open and when the mouse is released, the roller bottle will be placed in its holder.

5. Click on the buttons on the left side of the hybridization oven to set the temperature. Under these conditions, 47degreesC is an optimal temperature for prehybridization.

6. On the timer set the time for the prehybridization. One hour is usually sufficient. Click on GO.

7. When the prehybridization is complete, click on the hybridization oven to have the roller bottle return to the bench.

Task 4: Select a radioactively labeled probe.

1. Click the BACK button to move to the previous screen.

2. To select a radioactively labeled probe, click the arrow on the Plexiglass shield.

3. On the selection menu choose a cDNA or restriction fragment generated in other GDL modules. If you haven't performed the experiments in these modules you can choose one of the two premade probes.

4. If you make a mistake, drag the probe tube to the radioactive trash can.

Task 5: Denature the probe.

1. Drag the probe tube to the heat block. The DNA strands must be separated or denatured to allow the probe to bind to the DNA on the filter. Turn on the heat block. Set the temperature to 98degreesC.

2. Set the time of incubation for 5 minutes by clicking on the timer.

3. When the denaturation reaction is complete, drag the tube back to the probe rack.

Task 6: Add the probe to the roller bottle.

1. Use the pipettor to add 25 microliters of the radioactively labeled probe to the roller bottle with the filter.

2. Click on the NEXT button to move back to the hybridization oven.

3. Drag the capped roller bottle into the hybridization oven.

4. Set the temperature on the hybridization oven. The temperature determines the stringency of the hybridization. The optimal temperature is 42degreesC.

5. On the timer set the time for the hybridization. It must be at least for 4 hours.

6. Click here to view an animation of the hybridization process.

7. When the hybridization is complete, click on the hybridization oven door to bring the roller bottle back to the bench.

Task 7: Wash the filter.

1. To wash the filters, take off the cap, then drag the roller bottle to the radioactive waste bottle and pour off the hybridization solution.

2. Use the glass pipette to transfer 10 milliliters from the beaker with to the roller bottle.

3. Put on the cap and drag the roller bottle back into the hybridization oven.

4. Set the temperature on the hybridization oven. The wash temperature will also affect the stringency of the hybridization. A good wash temperature is 60degreesC.

5. On the timer set the time for the wash. One hour is usually sufficient. When the wash is complete, click on the door to bring the roller bottle to the bench.

Task 8: Place the filter in the X-ray cassette.

1. Click NEXT to move to the next screen.

2. Take the cap off of the roller bottle and use the forceps to drag the filter to the X-ray cassette.

3. Click the top of the X-ray cassette. After a few moments, you will see the filter exposing the X-ray film.

4. Take a picture of the filter by clicking on the film icon. The picture will be entered into your notebook.

5. Close the window to return to the bench.

Task 9: Interpret the results.

To determine the size of each band in each lane of the X-ray image follow the same procedure as in the Restriction Mapping module. The sizes of the bands on the filter are determined by comparing their location with the <u>gel marker</u> DNA fragments. The fragments of the gel marker are visible on the photograph of the filter taken when it was in the UV cross-linking oven.

1. Using the ruler, measure the distance from the starting well to each of these fragments and plot this information on a piece of graph paper or in a graphing program.

2. Make the Y axis the logarithm of fragment size and the X axis the distance migrated. Draw a straight line through the plotted points.

3. Determine the distance of each band on the X-ray image of the Southern blot from the well for that lane.

4. Plot each of these distances on the graph of the marker DNA size versus distance and determine the approximate size of each band.

5. You can now compare the size of the genomic DNA from a normal individual with the size of the DNA from patients with the diseases that you are investigating.

6. If the sizes are the same, what does this indicate? If the sizes are different, what might be the reasons for the difference in size?

If you already know the size of the cDNA you can compare the size of the band on the Southern blot with the size that you would predict from your knowledge of the cDNA size. Differences in size between the genomic DNA and cDNA are usually due to the presence of introns.

<u>TROUBLESHOOTING</u>

If you found no bands on the X-ray film the most likely causes are:
*   You didn't put the filter into the cross-linking oven.
*   You didn't cross-link the filter for the correct amount of time.
*   You forgot to put the probe into the roller bottle.
*   You didn't denature the probe in the heat block or you didn't set the temperature correctly on the heat block.
*   You didn't set the timer correctly for the denaturation process.
*   You forgot to add hybridization solution to the roller bottle.
*   You chose a cDNA colony from the cDNA cloning module to use as a probe that did not contain one of the two genes under study.
*   You chose a restriction fragment to use as a probe that did not contain at least part of one of the two genes under study.
*   You didn't leave the roller bottle in the hybridization oven long enough.
*   The temperature of the hybridization or wash was too high (higher than 70 degreesC).

If there are too many bands and the X-ray film is overly dark , in other words there is too much background - the most likely causes are:
*   The hybridization oven temperature was set too low for prehybridization or hybridization.
*   The wash temperature was too low.
*   You forgot to do the prehybridization.
*   You forgot to put the roller bottle in the hybridization oven after adding the probe.
*   You didn't hybridize for a sufficient length of time.
*   You forgot to wash the filters after hybridization.

EQUIPMENT AND REAGENT INFORMATION

Pipettor

Drag the automatic pipettor from its rack and a new disposable tip will appear. The tip will be automatically discarded after each use.

Click on the dark rectangular window on the pipettor to get a pop-up slider. Select the volume by adjusting the slider, by using either the sliding needle or the arrow buttons for fine control. Click OK or anywhere outside the slider panel to dismiss it.

Drag the pipettor to the rack containing tubes. When the tip of the pipettor is over a tube, the contents of the tube will be visible. This tells you which tube the pipettor tip is in.

When you release the mouse, if the volume is set within the correct limits, the reagent will be drawn up into the pipettor tip. If not, you will get an error message and will need to reset the volume.

Drag the pipettor to the tube to which you want to add the reagent. When the pipettor tip is over the tube, release the mouse. What was in the pipettor tip will now be added to that tube.

Heat Block

Turn on the heat block by clicking the ON/OFF switch. Click the control knob on the heater to set the temperature. A slider appears that will allow you to set the temperature.

If you push GO on the timer and you have forgotten to turn on the heat block you will get an error message.

Timer

When you click on the timer a slider will appear. Set the slider on the timer for the desired time. Click the GO button on the timer to start the timer. In GDL, time is speeded up, so the timed process will take only a few seconds.

If you need to adjust anything, click outside the time slider to dismiss it without starting the timer.

Gel Marker

The Gel Marker is DNA from bacteriophage lambda that has been cut with the restriction enzyme *Hind*III.

This marker has been used for many years because the resulting fragments produce a nice "ladder" of known sizes.

The sizes of the marker DNA from largest to smallest are: 23,130 basepairs (bp), 9416 bp, 6557 bp, 4361 bp, 2322 bp, 2027 bp, 564 bp, and 125 bp.

UV Cross-Linking Oven

To open or close the door of the UV crosslinking oven, click on the door. Turn on the UV crosslinking oven by clicking the ON/OFF switch.

Set the time of exposure by clicking on the keypad. A slider appears that will allow you to set the exposure time. To start the UV crosslinking push GO on the slider.

When the UV crosslinking oven is turned on and the time of exposure has been set, a new window appears to allow you to see the filter illuminated by the UV light.

If you click on the film icon in the window, a picture of the filter illuminated by UV light will appear in your notebook. You will want to refer to this picture when you interpret the results of your Northern blot.

Click in the upper left-hand box on the window to dismiss it. You can now open the oven door and drag the filter out with the forceps.

Forceps

To use the forceps, drag them toward the gel or filter.

Release the mouse when the tip of the forceps is over the filter. Now drag the forceps – the forceps will close on the filter and drag it.

Release the forceps when the filter is in the desired location.

Glass pipette

Drag the glass pipette from the bench.

Click on the volume element at the top of the glass pipette to get a pop-up slider.

Select the volume by adjusting the slide, by using either the sliding needle or the arrow buttons for fine control. Click OK or anywhere outside the slider panel to dismiss it.

Drag the glass pipette to a bottle or flask. When you release the mouse, if the volume is set within the correct limits, the reagent will be drawn up into the glass pipette. If not, you will get an error message and will need to reset the volume.

Drag the glass pipette to the tube or roller bottle to which you want to add the reagent. When the glass pipette is over the tube or roller bottle, release the mouse. The liquid in the glass pipette will now be added to the tube or roller bottle.

Hybridization solution

The hybridization solution for Southern blots has the following composition: 6x SSC, 5x Denhardt's reagent, 100 micrograms/milliliter salmon sperm DNA , 0.5% sodium dodecyl sulfate (SDS) and 50% formamide.

A stock solution of 2 x SSC is made by dissolving 175.3 grams of sodium chloride and 88.2 grams of sodium citrate in 800 ml of water. The pH is adjusted to 7.0 with the addition of sodium hydroxide, then the volume is adjusted to 1 liter.

A stock solution of 50x Denhardt's reagent is made by dissolving 5 grams of Ficoll, 5 grams of polyvinylpyrrolidone, and 5 grams of bovine serum albumin in 500 milliliters of water. Denhardt's reagent is used to reduce the effective aqueous volume of the hybridization solution.

Hybridization wash solution

The hybridization wash solution for high stringency washes of Southern blots has the following composition: 0.1x SSC, 0.1% SDS.

Roller bottle

Roller bottles are used because they allow hybridization to take place in a small volume while permitting the probe to contact the filter evenly over its entire surface.

To cap the roller bottle, drag the cap from the bench and release it over the roller bottle, it will automatically drop onto the roller bottle.

To pour off hybridization solution, drag the roller bottle to the liquid radioactive waste beaker. When the roller bottle is released, it will tip up and pour off the liquid.

Hybridization oven

Click on the hybridization oven door to open or close it. Set the oven temperature by clicking on the buttons to the left of the door.

Turn on the oven by clicking the ON/OFF switch. Set the timer. The timer will not run if the oven is not turned on.

X-ray cassette

The forceps must be used to transfer the filter from the roller bottle to the X-ray cassette.

To activate the X-ray cassette, click on the top part of the cassette.

Northern Blot

The procedures in this lab will enable you to:

- **Choose genomic RNAs for the Northern blot.**
- **Select a fragment of DNA to use as a probe.**
- **Hybridize the Northern blot with the probe.**
- **Interpret the hybridization pattern.**

PURPOSE

Northern blotting is used to determine the size of RNAs that hybridize to specific probes.

You can use Northern blotting to compare the RNA from normal individuals with the RNA from patients with the diseases under investigation.

To get started immediately, go to Task 1.

TABLE OF CONTENTS

Background
Introduction
Conceptual and technical overview
Task 1: Choose RNAs to fill the lanes in the gel.
Task 2: Cross-link RNA to the filter.
Task 3: Set up the prehybridization.
Task 4: Select a radioactively labeled probe.
Task 5: Denature the probe.
Task 6: Add probe to the roller bottle.
Task 7: Wash the filter.
Task 8: Place the filter in the X-ray cassette.
Task 9: Interpret the results.
Troubleshooting
Equipment and reagent information

BACKGROUND

Northern blotting is used to determine whether a particular RNA is present in a tissue or whether the RNA is of the appropriate size. The procedure for separating RNA by gel electrophoresis and transferring it to a filter is very similar to that used for DNA. DNA blotting was developed by E. M. Southern and is called Southern blotting; by analogy, RNA blotting is called Northern blotting. The developer of Northern blotting, Jim Alewine, thus never achieved the notoriety of E. M. Southern.

The process of a probe finding an RNA that it is homologous to is called hybridization. For hybridization to occur, the probe has to be single-stranded, and the RNA should be free of internal base-pairing which causes the formation of stem/loop structures. When a probe is made from double-stranded DNA, the bonds that hold the two strands together have to be broken. This process is known as denaturing the DNA. When the probe finds its homologous RNA, the two strands will zip up to form a RNA/DNA hybrid.

Task 1, 2, 3, 4, 5, 6, 7, 8, 9

APPLICATIONS

When RNA on a Northern blot is probed with fragments of a genomic DNA clone, the effects of mRNA processing can be detected. One of the major steps in mRNA processing is the removal of introns by splicing. Genomic DNA represents the unspliced form while the mRNA is the spliced form.

Northern blotting can be used to identify certain types of mutations in disease genes. Most point mutations in the coding region of genes do not have an effect on the amount or length of the mRNA made from the gene. Thus, these mutations cannot be detected by Northern blotting. However, deletions, insertions and translocations usually do have an effect on mRNA amount, size or both, and thus can be detected by Northern blotting.

Task 1, 2, 3, 4, 5, 6, 7, 8, 9

INTRODUCTION

Differences in the abundance and/or the size of RNA molecules can provide important information as to the genetic defects responsible for a disease.

To make a Northern blot, RNA is isolated from tissue, then separated by size using agarose gel electrophoresis. The separation of RNA molecules by agarose gel electrophoresis is similar to the separation of DNA fragments performed in the Restriction Mapping module.

When the RNA fragments are sufficiently separated, the electrophoresis is stopped and the gel is placed on an apparatus that uses capillary action to transfer the RNA from the gel to a nylon filter. Essentially the same apparatus is used for Northern blotting and Southern blotting. Click here to view a movie that shows how this blotting apparatus is set up.

After the transfer is complete, the gel and nylon filter are placed on the bench and the position of the gel wells are marked on the filter. In this module you will find the gel and nylon filter already placed on the bench with the wells marked.

Task 1, 2, 3, 4, 5, 6, 7, 8, 9

CONCEPTUAL AND TECHNICAL OVERVIEW

RNA Extraction and Separation

RNA for Northern blotting can come from a variety of sources. Each type of tissue in the body has a different assortment of RNAs for making the proteins for that tissue's specialized functions. Therefore, your choice of tissue, from which the RNA will be extracted, is an important decision and should be made carefully. Knowing where a particular protein is found, based on the gene under investigation, can help you make this decision.

- Beta-globin protein is made in relatively high amounts in the spleen and to a lesser extent in the liver. Very little (if any) globin is made in the skin.

- p53 protein is found in low amounts in almost all tissues. Increased amounts of p53 are found in tissues treated with ionizing radiation, those that have undergone DNA damage or oxygen deprivation and in some tumors.

After extraction and purification, the RNA is separated by agarose gel electrophoresis. In GDL, approximately 10 micrograms of total RNA is loaded in each gel. Because RNAs are made in different sizes, there is no need to cleave them first, as is done with restriction enzymes for DNA. On the other hand, RNA molecules tend to fold and base-pair to themselves, creating "secondary structure." To prevent secondary structure formation during gel electrophoresis, the gel is made with a buffer that diminishes the likelihood that the RNA will bind to itself.

Task 1, 2, 3, 4, 5, 6, 7, 8, 9

Probe

Detecting a specific RNA species among the RNAs on the Northern blot requires a probe, which is usually a fragment of DNA containing a sequence from the gene of interest.

In GDL, you can make probes from DNAs that you have generated in other modules, or you can use one of the "premade" probes. If you select a DNA from another module, you need to make sure that the DNA that you choose actually contains a part of the gene that you are interested in.

If you choose a colony containing a cDNA clone from the cDNA Cloning module to use as a probe, be sure that there is a positive colony in the same place on both of the replica filters. If not, you may be selecting a background colony that will lack a cDNA from one of the genes under investigation.

If you choose a restriction fragment from the Restriction Mapping module to use as a probe, be sure that the fragment contains a part of the gene under investigation. Some of the fragments will contain only a vector sequence and these will not hybridize to any RNAs on the Northern blot.

Before you can use a fragment of DNA as a probe, the two strands need to be separated. This is necessary because only single-stranded DNA can hybridize effectively to the RNA on the Northern blot.

Separating the DNA strands involves breaking the hydrogen bonds that hold the bases together a process known as denaturation. Denaturation can be achieved either chemically or by heating the DNA. The temperature at which the DNA strands separate is referred to as its "melting point."

Probe denaturation is frequently achieved by heating the DNA to about 100 degrees Celsius for a few minutes.

Task 1, 2, 3, 4, 5, 6, 7, 8, 9

Radioactive Labeling

To detect successful binding of the probe to the RNA on the Northern blot requires putting a label on the probe. One of the most widely used labels in molecular biology is radioactivity. This is usually in the form of a radioactive atom incorporated into a biologically active molecule.

For hybridization, a radioactive variant of phosphorous, ^{32}P, is frequently incorporated into the probe. When ^{32}P decays it emits a beta particle that can be detected on X-ray film.

To incorporate ^{32}P into a DNA fragment for Northern blot hybridization, a commonly used method is nick translation.

Nick translation involves the introduction of a break, or a "nick," into one strand of the DNA molecule. Next, DNA polymerase is added to begin priming at the site of the nick and to elongate the DNA molecule, adding nucleotides in the process.

One of the four nucleotides, frequently dCTP, has a ^{32}P molecule incorporated into it. Whenever dCTP is added to the growing DNA strand its radioactive label will be included in the DNA.

As the polymerase moves down the complementary strand it removes the old DNA that was there prior to the introduction of the nick. Conceptually then, the break in the DNA is moved down the DNA with the polymerase. The nick is translated (or moved) down the DNA. Click here to view an animation of nick translation.

Although the amounts and intensities of radioactive compounds used in most molecular biology procedures are quite small compared to medical or industrial procedures, safety precautions must be exercised. In GDL, of course there is no real risk, but to encourage good habits, safe practices for handling radioactive materials are included.

Because Plexiglass absorbs radioactive beta emissions, the labeled probe is kept behind a Plexiglass shield. Any materials that have been in contact with radioactivity should be disposed of in the radioactive trash container.

Task 1, 2, 3, 4, 5, 6, 7, 8, 9

UV Cross-Linking of RNA to Nylon Filters

Permanent attachment of RNA to the nylon filter requires exposing it to ultraviolet (UV) light.

At the intensities of UV light normally used, very short (~ 10 second) exposure to the light is sufficient to cause the RNA to become attached or cross-linked to the nylon filter.

During gel electrophoresis, the RNA is bound by ethidium bromide. When exposed to UV light, ethidium bromide emits an orange color, making RNA visible on the nylon filter.

Task 1, 2, 3, 4, 5, 6, 7, 8, 9

Hybridization

The process of allowing one strand of a nucleic acid (DNA or RNA) to find and base-pair with its complementary strand is called hybridization.

Nucleic acids also have a tendency to bind to other nucleic acids to which they are only partially complementary. In this case some or most of the bases don't match the bases opposite them. This is called non-specific hybridization or background hybridization.

It is challenging to get hybridization to occur within a reasonable time frame and with a minimum of background hybridization. Specific hybridization requires a certain set of conditions, beginning with the use of a special solution.

This special hybridization solution generally contains large amounts of nonlabeled RNA or DNA whose role is to bind to all molecules that might nonspecifically bind nucleic acids, before the labeled probe binds to these molecules. The hybridization solution usually also contains a substance that reduces the effective volume in which the hybridization takes place.

Two variables can be easily manipulated to alter the hybridization conditions - salt concentration and temperature.

A dynamic tension exists between the hydrogen bonds that hold two complementary nucleic acid strands together and their negatively charged phosphate backbones which tend to push them apart. Raising the salt concentration of the hybridization solution lessens the repulsive force of the phosphates. Although this stabilizes the base-pairing, it also allows more mismatched base-pairing to occur.

To reduce the number of variables, in GDL the hybridization solution contains an optimal salt concentration that reduces the amount of mismatched base-pairing.

For a particular salt concentration, changing the temperature will alter the ratio of correct hybridization to mismatched base-pairing. As the temperature is raised, mismatched base-pairs will tend to pull apart, because they have weak or no hydrogen bonds holding them together. The converse is also true: As the temperature is lowered, more mismatching will occur.

Of course, there is an upper limit to the temperature that can be used. Above a certain temperature, even the bonds between correctly paired bases will be broken. The salt and temperature of a hybridization define the stringency conditions of the hybridization. Low salt and high temperature are high stringency conditions, high salt and low temperature are low stringency conditions.

Task 1, 2, 3, 4, 5, 6, 7, 8, 9

Temperature and Time of Hybridization

For the hybridization solution used in GDL, a good hybridization temperature for Northern blots is 42 degreesC. If hybridization is carried out at lower temperatures, there will be more background.

The time period of hybridization is not critical beyond a minimum length. Adequate time is needed for the probe to find its complementary strand. This depends on several factors, including the abundance of the complementary strand, and the amount of probe.

Hybridizations are usually performed overnight. In GDL the minimum hybridization time is 4 hours.

Task 1, 2, 3, 4, 5, 6, 7, 8, 9

Prehybridization

The purpose of prehybridization is to prepare the filters for hybridization with radioactive probe. This process begins with exposing the filters to an abundance of DNA or RNA, thereby reducing the likelihood that the probe will bind to the wrong places.

Prehybridization is performed by adding hybridization solution to the filters and incubating them in the hybridization oven before adding the radioactively labeled probe.

To be effective, prehybridization should be performed at a temperature that is at least as high as the hybridization temperature. Prehybridization is often performed at 5 degrees warmer than the hybridization temperature.

Prehybridization is normally performed for at least 1 hour.

Task 1, 2, 3, 4, 5, 6, 7, 8, 9

Filter Washing

Washing unbound probe from filters after hybridization is an essential step to reduce background.

The first step is to pour off the hybridization solution from the roller bottle. Because the solution also contains radioactively labeled probe, it must be poured into the liquid radioactive waste container.

Effective washing of filters depends on the same variables as hybridization: salt concentration and temperature. In GDL, the salt concentration of the wash solution has already been optimized.

The temperature of washing depends on the desired stringency. For a high-stringency wash, the temperature should be 60degreesC.

Filters should remain in the wash solution for 30 minutes.

If the wash temperature is above 70 degrees the probe will become unbound to the RNA and you will not see anything on the X-ray film.

If the wash temperature is below 32 degrees, the background will be excessively strong.

Task 1, 2, 3, 4, 5, 6, 7, 8, 9

PROCEDURES

Task 1: Choose RNAs to fill the lanes in the gel.

1. On the lab bench, you will find a gel on top of a filter. This is what the gel looks like after the process of transferring the RNA to the filter is completed. Click here to view a movie on the RNA transfer process.

2. To choose the RNAs for each lane of the Northern blot click the arrowhead next to the gel to get the RNA selection menu. The first lane of the gel (the one at the far left) contains gel marker RNA.

3. Continue to select RNAs until you have chosen all that are relevant to the disease that you are investigating or until you have filled all the lanes. The contents of each lane will be automatically entered into your notebook

Task 2: Cross-link RNA to the filter.

1. Use the forceps to peel away the gel by dragging it from the raised corner. Drag it to the trash can.

42

2. To <u>fix the RNA</u> to the nylon filter you must place it under ultraviolet (UV) light. Open the <u>UV cross-linking oven</u> by clicking on the door.

3. Use the forceps to drag the filter to the oven.

4. Close the oven door by clicking on it.

5. Click on the oven keypad and use the slider to set the time of UV exposure for 10 seconds.

6. Turn on the oven by clicking the ON/OFF switch. A new window will open to give you a good view of the filter inside the oven.

7. Click on the film icon at the bottom of the window to take a picture of the <u>RNA bound to the filter</u>. The picture can be accessed from your notebook.

8. Close the window to return to the bench.

Task 3: <u>Set up the prehybridization.</u>

1. Use the forceps to drag the filter from the UV cross-linking oven and into the <u>roller bottle</u>. Place the top on the roller bottle.

2. Click the NEXT button twice to move to the screen with the hybridization oven.

3. Use the <u>glass pipette</u> to transfer 10 ml of <u>hybridization solution</u> from its flask to the roller bottle. Cap the roller bottle.

4. To <u>prehybridize</u> the filter, drag the roller bottle to the <u>hybridization oven</u>. The door will automatically open and when the mouse is released, the roller bottle will be placed in its holder.

5. Click on the buttons on the left side of the hybridization oven to set the <u>temperature</u>. Under these conditions, 47degreesC is an optimal temperature for prehybridization.

6. On the timer <u>set the time for the prehybridization</u>. One hour is usually sufficient. Click on GO.

7. When the prehybridization is complete, click on the hybridization oven to have the roller bottle return to the bench.

Task 4: <u>Select a radioactively labeled probe.</u>

1. Click the BACK button to move to the previous screen.

2. To select a <u>radioactively</u> labeled <u>probe</u>, click the arrow on the Plexiglass shield.

3. On the selection menu choose a cDNA or restriction fragment generated in other GDL modules. If you haven't performed the experiments in these modules you can choose one of the two premade probes.

4. If you make a mistake, drag the probe tube to the radioactive trash can.

Task 5: <u>Denature the probe.</u>

1. Drag the probe tube to the <u>heat block</u>. The DNA strands must be separated or denatured to allow the probe to bind to the RNA on the filter. Turn on the heat block. Set the temperature to 98degreesC.

2. Set the time of incubation for 5 minutes by clicking on the <u>timer</u>.

3. When the denaturation reaction is complete, drag the tube back to the probe rack.

Task 6: <u>Add the probe to the roller bottle</u>.

1. Use the pipettor to add 25 microliters of the radioactively labeled probe to the roller bottle with the filter.

2. Click on the NEXT button to move back to the hybridization oven.

3. Drag the capped roller bottle into the hybridization oven.

4. Set the temperature on the hybridization oven. The temperature determines the <u>stringency</u> of the hybridization. The optimal temperature is 42degreesC.

5. On the timer set the <u>time</u> for the hybridization. It must be at least for 4 hours.

6. <u>Click here to view an animation of the hybridization process</u>.

7. When the hybridization is complete, click on the hybridization oven door to bring the roller bottle back to the bench.

Task 7: <u>Wash the filter</u>.

1. To <u>wash the filters</u>, take off the cap, then drag the roller bottle to the radioactive waste bottle and pour off the hybridization solution.

2. Use the glass pipette to transfer 10 milliliters from the beaker with to the roller bottle.

3. Put on the cap and drag the roller bottle back into the hybridization oven.

4. Set the temperature on the hybridization oven. The wash temperature will also affect the <u>stringency</u> of the hybridization. A good wash temperature is 60degreesC.

5. On the timer set the <u>time</u> for the wash. One hour is usually sufficient. When the wash is complete, click on the door to bring the roller bottle to the bench.

Task 8: <u>Place the filter in the X-ray cassette</u>.

1. Click NEXT to move to the next screen.

2. Take the cap off of the roller bottle and use the forceps to drag the filter to the <u>X-ray cassette</u>.

3. Click the top of the X-ray cassette. After a few moments, you will see the filter exposing the X-ray film.

4. Take a picture of the filter by clicking on the film icon. The picture will be entered into your notebook.

5. Close the window to return to the bench.

Task 9: <u>Interpret the results</u>.

To determine the size of each band in each lane of the X-ray image follow the same procedure as in the Restriction Mapping module. The sizes of the bands on the filter are determined by comparing their location with the <u>gel marker</u> RNA fragments. The fragments of the gel marker are visible on the photograph of the filter taken when it was in the UV cross-linking oven.

1. Using the ruler, measure the distance from the starting well to each fragment and plot this information on a piece of graph paper or in a graphing program.

2. Make the Y axis the logarithm of fragment size and the X axis the distance migrated. Draw a straight line through the plotted points.

3. Determine the distance of each band on the X-ray image of the Northern blot from the well for that lane.

4. Plot each of these distances on the graph of the marker RNA size versus distance and determine the approximate size of each band.

5. Compare the size and amount of the RNA from a normal individual with the size and amount of RNA from patients with the disease that you are investigating. If the sizes and amounts are the same, what does this indicate? If the sizes or amounts are different, what might be the reasons for this?

6. If you know the cDNA sequence compare the size of the band on the Northern blot with the size that you would predict from your knowledge of the cDNA sequence.

TROUBLESHOOTING

If you found no bands on the X-ray film the most likely causes are:
- You didn't put the filter into the cross-linking oven.
- You didn't cross-link the filter for the correct amount of time.
- You forgot to put the probe into the roller bottle.
- You didn't denature the probe in the heat block or you didn't set the temperature correctly on the heat block.
- You didn't set the timer correctly for the denaturation process.
- You forgot to add hybridization solution to the roller bottle.
- You chose a cDNA colony from the cDNA cloning module to use as a probe that did not contain one of the two genes under study.
- You chose a restriction fragment to use as a probe which did not contain at least part of one of the two genes under study.
- You didn't leave the roller bottle in the hybridization oven long enough.
- The temperature of the hybridization or wash was too high (higher than 70 degreesC).

If there are too many bands and the X-ray film is overly dark, in other words there is too much background - the most likely causes are:

- The hybridization oven temperature was set too low for prehybridization or hybridization.
- The wash temperature was too low.
- You forgot to do the prehybridization.
- You forgot to put the roller bottle in the hybridization oven after adding the probe.
- You didn't hybridize for a sufficient length of time.
- You forgot to wash the filters after hybridization.

EQUIPMENT AND REAGENT INFORMATION

Pipettor

Drag the automatic pipettor from its rack and a new disposable tip will appear. The tip will be automatically discarded after each use.

Click on the dark rectangular window on the pipettor to get a pop-up slider. Select the volume by adjusting the slider, by using either the sliding needle or the arrow buttons for fine control. Click OK or anywhere outside the slider panel to dismiss it.

Drag the pipettor to the rack containing tubes. When the tip of the pipettor is over a tube, the contents of the tube will be visible. This tells you which tube the pipettor tip is in.

When you release the mouse, if the volume is set within the correct limits, the reagent will be drawn up into the pipettor tip. If not, you will get an error message and will need to reset the volume.

Drag the pipettor to the tube to which you want to add the reagent. When the pipettor tip is over the tube, release the mouse. What was in the pipettor tip will now be added to that tube.

Heat Block

Turn on the heat block by clicking the ON/OFF switch. Click the control knob on the heater to set the temperature. A slider appears that will allow you to set the temperature.

If you push GO on the timer and you have forgotten to turn on the heat block you will get an error message.

Timer

When you click on the timer a slider will appear. Set the slider on the timer for the desired time. Click the GO button on the timer to start the timer. In GDL, time is speeded up, so the timed process will take only a few seconds.

If you need to adjust anything, click outside the time slider to dismiss it without starting the timer.

Gel Marker

The Gel Marker consists of a mixture of RNA molecules of known size. The sizes of this gel marker in bases are: 9490; 7460; 4400; 2370; 1350; 240.

UV Cross-Linking Oven

To open or close the door of the UV crosslinking oven, click on the door. Turn on the UV crosslinking oven by clicking the ON/OFF switch.

Set the time of exposure by clicking on the keypad. A slider appears that will allow you to set the exposure time. To start the UV crosslinking push GO on the slider.

When the UV crosslinking oven is turned on and the time of exposure has been set, a new window appears to allow you to see the filter illuminated by the UV light.

If you click on the film icon in the window, a picture of the filter illuminated by UV light will appear in your notebook. You will want to refer to this picture when you interpret the results of your Northern blot.

Click in the upper left-hand box on the window to dismiss it. You can now open the oven door and drag the filter out with the forceps.

Forceps

To use the forceps, drag them toward the gel or filter.

Release the mouse when the tip of the forceps is over the filter. Now drag the forceps – the forceps will close on the filter and drag it.

Release the forceps when the filter is in the desired location.

Glass Pipette

Drag the glass pipette from the bench.

Click on the volume element at the top of the glass pipette to get a pop-up slider.

Select the volume by adjusting the slide, by using either the sliding needle or the arrow buttons for fine control. Click OK or anywhere outside the slider panel to dismiss it.

Drag the glass pipette to a bottle or flask. When you release the mouse, if the volume is set within the correct limits, the reagent will be drawn up into the glass pipette. If not, you will get an error message and will need to reset the volume.

Drag the glass pipette to the tube or roller bottle to which you want to add the reagent. When the glass pipette is over the tube or roller bottle, release the mouse. The liquid in the glass pipette will now be added to the tube or roller bottle.

Hybridization Solution

The hybridization solution for Northern blots has the following composition: 6x SSC, 5x Denhardt's reagent, 100 micrograms/milliliter salmon sperm DNA , 0.5% sodium dodecyl sulfate (SDS) and 50% formamide.

A stock solution of 20x SSC is made by dissolving 175.3 grams of sodium chloride and 88.2 grams of sodium citrate in 800 ml of water. The pH is adjusted to 7.0 with the addition of sodium hydroxide, then the volume is adjusted to 1 liter.

A stock solution of 50xDenhardt's reagent is made by dissolving 5 grams of Ficoll, 5 grams of polyvinylpyrrolidone, and 5 grams of bovine serum albumin in 500 ml of water. Denhardt's reagent is used to reduce the effective aqueous volume of the hybridization solution.

Hybridization Wash Solution

The hybridization wash solution for high stringency washes of Northern blots has the following composition: 0.1x SSC, 0.1% SDS.

Roller Bottle

Roller bottles are used because they allow hybridization to take place in a small volume while permitting the probe to contact the filter evenly over its entire surface.

To cap the roller bottle, drag the cap from the bench and release it over the roller bottle, it will automatically drop onto the roller bottle.

To pour off hybridization solution, drag the roller bottle to the liquid radioactive waste beaker. When the roller bottle is released, it will tip up and pour off the liquid.

Hybridization oven

Click on the hybridization oven door to open or close it. Set the oven temperature by clicking on the buttons to the left of the door.

Turn on the oven by clicking the ON/OFF switch. Set the timer. The timer will not run if the oven is not turned on.

X-Ray Cassette

The forceps must be used to transfer the filter from the roller bottle to the X-ray cassette.

To activate the X-ray cassette, click on the top part of the cassette.

Subcloning Module

OBJECTIVES

The procedures in this lab will enable you to:

- **Ligate DNA fragments into a cloning vector.**
- **Transform bacteria with the ligated DNA.**
- **Use blue/white screening to identify colonies with subclones.**

PURPOSE

Subcloning is used to insert fragments of DNA into a cloning vector.

You can subclone DNA fragments produced in the Restriction Mapping module that you want to sequence in the DNA Sequencing module or make recombinant protein in the Protein Expression module.

To get started immediately, go to Task 1.

TABLE OF CONTENTS

Background
Introduction
Conceptual and technical overview
Task 1: Select the DNA fragments to be subcloned.
Task 2: Select the cloning vector.
Task 3: Set up the ligation.
Task 4: Incubate the ligation reaction.
Task 5: Transform bacteria with the ligated DNA.
Task 6: Plate the transformed bacteria.
Task 7: Incubate the transformed bacteria.
Task 8: Determine the cloning efficiency.
Troubleshooting
Equipment and reagent information

BACKGROUND

One of the major achievements of molecular biology is the ability to express foreign DNA in bacteria. This procedure allows for the amplification of large amounts of DNA from any source, a process which is called cloning.

Cloning entails three main steps. The first step is cutting DNA at specific places by using restriction enzymes. The second step is covalently linking pieces of DNA by using the enzyme ligase. The third step is the engineering of circular DNA molecules containing the signals required to replicate themselves in bacteria and incorporating a means of selecting for their presence in bacteria. These are called plasmid cloning vectors; they have a bacterial origin of replication and contain genes that confer resistance to one or more antibiotics.

APPLICATIONS

Genetic engineering has many industrial applications. For example, detergents are now made with enzymes that have been modified to make them better able to degrade proteins. Bacteria and fungi have been genetically engineered for use in waste treatment and hazardous waste cleanup. Plants have been modified for use by mining companies to grow on mine spoils which otherwise would remain barren.

Subcloning is also the first step in the production of vectors for gene therapy and is used to insert genes into plants for agricultural improvement.

Task 1, 2, 3, 4, 5, 6, 7, 8

INTRODUCTION

Subcloning is used for many purposes, including generating sufficient quantities of DNA for sequencing and making recombinant protein for antibody production.

To make a subclone, DNA fragments are ligated into a cloning vector and the resulting plasmid is mixed with bacteria. Included in the cloning vector is a gene that confers resistance to an antibiotic. Antibiotic resistance is used to select for bacteria that have successfully taken up the plasmid DNA.

After the plasmid DNA is taken up by the bacteria , in a process known as transformation, the bacteria are spread (or "plated") on medium that contains the antibiotic. Only bacteria that have the cloning vector survive. Some cloning vectors also have a means of telling whether the DNA fragment has been inserted into them. One of these is known as blue-white screening.

To make recombinant protein, an expression cloning vector is used. In these vectors, the goal is to fuse the protein you wish to express with a bacterial protein that is easy to purify. For this to work, the open reading frame of the inserted fragment must be in the same reading frame as the protein to which it is being fused.

Task 1, 2, 3, 4, 5, 6, 7, 8

CONCEPTUAL AND TECHNICAL OVERVIEW

DNA Fragments for Subcloning

DNA fragments used for subcloning can come from restriction digests or PCR amplification. It is useful to purify the DNA fragment to be subcloned away from other DNA fragments that are not to be subcloned.

One way to do this is by first separating DNA fragments by agarose gel electrophoresis, and then cutting out the region of the gel that contains the DNA fragment you want to subclone. The DNA within this small piece of agarose is then extracted from the gel material.

One means of extracting the DNA is to melt the agarose and allow the DNA to bind to small glass beads. The DNA stuck to the beads is then washed. Finally, the DNA is removed from the beads by changing the salt concentration of the solution. Click here to view a movie that describes the isolation of DNA fragments from an agarose gel.

Task 1, 2, 3, 4, 5, 6, 7, 8

Inserting DNA Fragments into Cloning Vectors

DNA fragments are usually inserted into cloning vectors through use of restriction enzymes. When restriction enzymes cut DNA, they often leave a small region of single-stranded DNA known as a sticky end. Two pieces of DNA that have been cut with the same restriction enzyme will have cohesive sticky ends.

The cohesive sticky ends allow the two pieces of DNA to bind together through DNA base-pairing. This, in turn, stabilizes the junction until the enzyme ligase can create a covalent bond between the two pieces of DNA.

Different enzymes recognize different sequences, or restriction enzyme sites, on the DNA, and will therefore cut to yield different sticky ends. The sticky ends produced by different enzymes are not cohesive; they are not complementary and cannot base-pair with each other. Therefore, it is essential to use the same restriction enzymes to cut the cloning vector as you used to cut the DNA fragment you wish to insert into it.

Some restriction enzymes leave flush or blunt ends instead of single-stranded sticky ends. Ligation of two blunt-ended DNAs is much more difficult than ligation of two DNAs with complementary sticky ends.

If a fragment of DNA has the same restriction sites at both its ends, it can be inserted into a cloning vector that has also been cut with the same restriction enzyme. In this case, there are two potential problems:

- The piece of DNA can be inserted in either orientation.
- The cloning vector can ligate back to itself without inserting the fragment of DNA.

To sequence DNA, the orientation is usually not critical when inserting the DNA into the cloning vector. However, for expression cloning, insertion in the wrong orientation will result in no recombinant protein.

When you use cloning vectors in GDL that have two cohesive sticky ends, they will be treated in such a way as to prevent them from ligating back to themselves. This treatment uses the enzyme alkaline phosphatase, which removes a phosphate from the end of the cloning vector DNA.

Task 1, 2, 3, 4, 5, 6, 7, 8

Cloning Vectors

Plasmid cloning vectors contain several features:

- An origin of replication, which allows the plasmid to be replicated inside a bacterial cell.

- At least one selectable marker that confers resistance to an antibiotic so you can select the bacteria that have taken up the plasmid. Any bacteria that do not contain the vector will die in the presence of antibiotic.

- One or more restriction sites that occur only once on the plasmid. These allow you to insert foreign DNA into the plasmid without cutting something important (like the origin of replication).

- A means of identifying which plasmids have incorporated foreign DNA; in other words, a means of distinguishing between recombinant plasmids and nonrecombinant plasmids.

Task 1, 2, 3, 4, 5, 6, 7, 8

pUC19

In the pUC19 plasmid, the origin of replication is marked ori. Click here to view a detailed map of the pUC19 plasmid.

The selectable marker is a gene that confers resistance to the antibiotic ampicillin. This marker is on the part of the plasmid labeled amp^r.

Thirteen unique restriction sites are located close together in a region known as the polylinker. In addition, there are several other unique restriction enzyme sites at other locations on the pUC19 plasmid. Recombinant plasmids are identified by blue/white screening.

Task 1, 2, 3, 4, 5, 6, 7, 8

Expression Cloning Vector, pMAL

The expression cloning vector used in GDL is the plasmid pMAL. Successful subcloning into pMAL results in a fusion protein consisting of the *E. coli* maltose binding protein fused to the inserted protein. Click here to view a detailed map of the pMAL plasmid.

To make a fusion protein, the open reading frame (ORF) of the inserted DNA must be in the same reading frame as that of malE, which codes for maltose binding protein. Because of the high affinity of maltose binding protein for maltose, fusion proteins can be purified on beads that have maltose attached to them.

In'the pMAL plasmid, the origin of plasmid replication is marked ori. The selectable marker is resistance to ampicillin and is labeled amp^r. Seven restriction sites only found once on the plasmid are close together in a region known as the polylinker. Foreign DNA is typically inserted into the polylinker. Recombinant plasmids are identified by blue/white screening.

Task 1, 2, 3, 4, 5, 6, 7, 8

Ligation of DNA

Successful ligation of fragment DNA into a cloning vector depends on two variables:

- The two pieces of DNA to be ligated must have complementary ends, either blunt or sticky.

- The ratio of DNA fragment to cloning vector must be correct. The best ratio is usually a 1:1 molar ratio. This means that there is one molecule of DNA fragment for each molecule of cloning vector.

A quick way to determine the approximate molar ratio is by first comparing the length of the DNA fragment to the length of the cloning vector. For example, if the DNA fragment is 600 bp and the cloning vector is 3000 bp, you will need five times the amount of vector compared to fragment (based on molecular weight) to obtain an equal number of DNA fragment and vector molecules. Thus, you would combine in your ligation reaction 1 microgram of cloning vector and 200 nanograms of DNA fragment.

The enzyme ligase is purified from bacteria infected with the T4 virus. It catalyzes the formation of a phosphodiester bond between the 5' phosphate on one DNA and the 3' hydroxyl on a juxtaposed DNA. When DNA is cut by a restriction enzyme, this phosphodiester bond is cleaved. Thus, ligase is effective in resealing two pieces of cut DNA.

Ligase requires a source of energy to make the covalent bond between two pieces of DNA. This is provided by adenosine triphosphate (ATP), which is added to the ligation reaction. The optimal temperature for ligation is 16 degrees C.

Task 1, 2, 3, 4, 5, 6, 7, 8

Transformation of Bacteria with DNA

Bacteria can take up DNA, a process known as transformation. Most molecular biology experiments use strains of the'bacterium- E. coli, which is found in the human intestine. The efficiency of transformation can be greatly increased by pretreating the bacteria with substances that increase the permeability of their membranes. Once treated in this way, the bacteria are referred to as competent cells.

The first step of the transformation process is to mix the DNA with the competent cells on ice. It has been found that a brief exposure to higher temperature will also increase the efficiency of transformation. This is known as a heat shock. Subsequent to the heat shock, the bacteria are again placed on ice to allow them to recover.

Then growth medium (Luria broth) is added and the bacteria are incubated at 37 degrees C for 1 hour. This allows the bacteria that have taken up the cloning vector to begin to express the antibiotic resistance gene.

Task 1, 2, 3, 4, 5, 6, 7, 8

Plating of Transformed Bacteria

Selection of bacteria that have taken up a cloning vector is usually performed in petri dishes containing a combination of Luria broth and agar (a substance with the consistency of hardened gelatin). The Luria broth provides all the nutrients needed for bacterial growth, and the agar provides a semisolid, absorbent surface. Also included is the antibiotic for selecting transformed bacteria. The petri dish with nutrient agar medium is commonly referred to as a bacterial plate.

After transformation, the bacteria are spread on the bacterial plate; this procedure is known as plating the bacteria. The agar absorbs the liquid in which the bacteria have been growing, leaving the bacteria on the surface of the plate. Successful plating allows space between individual bacterial cells on the plate. When the plates are incubated at 37 degrees C, the bacteria divide many times, making visible colonies. Each colony is formed by cells descended from a single bacterial cell.

Task 1, 2, 3, 4, 5, 6, 7, 8

Blue/White Screening

Several cloning vectors use a change in color of the bacterial colony to indicate that a DNA fragment has been successfully inserted into them.

Both pUC19 and pMAL use modifications of the lactose operon for this purpose. The lactose operon includes a series of bacterial genes that respond to the sugar lactose. One of these genes is lacZ, which codes for beta-galactosidase.

Included in the cloning vector is a truncated portion of lacZ known as lacZ', which encodes the first 146 amino acids of beta-galactosidase. The host bacterium has a mutated form of lacZ which, on its own, is not functional. However, when the cloning vector makes the lacZ' fragment, this can combine with the host bacterium's mutated form of lacZ to make a functional enzyme.

Upstream (in the 5' direction) of the lacZ' coding region are the control sequences of the lacZ operon. These sequences control whether the genes in the operon are expressed or repressed (not expressed). In the absence of lactose in the bacterial growth medium, expression is repressed. When lactose, or IPTG (an analog of lactose), is added to the medium, the repression is released and high-level expression occurs.

X-gal (5-bromo-4-chloro-3-indole-beta-D-galactoside) is a substrate of beta-galactosidase. Normally colorless, it turns blue when cleaved by beta-galactosidase. Thus, when bacteria containing an intact cloning vector are plated on medium containing IPTG and X-gal lacZ is no longer repressed, and the colonies turn blue.

In pUC 19, the polylinker is inserted 5 amino acids downstream of the start codon for lacZ'. Because the polylinker consists of exactly 54 base-pairs, (containing the recognition sequences for 13 restriction enzymes), this region codes for 18 amino acids. The addition of these 18 amino acids has no effect on the function of the lacZ' protein. However, when a DNA fragment is inserted into the polylinker, this disrupts the coding sequence of lacZ', and no functional beta-galactosidase is made. Thus, on plates containing IPTG and X-gal, bacteria that contain DNA fragments inserted into the pUC19 polylinker are white.

In the expression vector pMAL, the lacZ' coding sequence comes immediately after the malE coding sequence. When the cloning vector is intact (with no inserted DNA), a fusion protein is made in which the first half is maltose binding protein and the second half is the beta-galactosidase fragment.

Expression of this protein is controlled by similar control sequences to those found on pUC19. Thus, on IPTG and X-gal plates, bacteria that contain intact pMAL are blue. When a DNA fragment is inserted into the pMAL polylinker, it disrupts the fusion to lacZ'. Thus, on plates containing IPTG and X-gal, bacteria that contain DNA fragments ligated into pMAL are white.

It is always a good idea to include a control to make sure everything is working as it should. A good control to include is transformation with the intact cloning vector.

Task 1, 2, 3, 4, 5, 6, 7, 8

PROCEDURES

Task 1: <u>Select the DNA fragments to be subcloned.</u>

1. Click the arrow on the fragments rack to get the DNA fragment selection menu.

2. To select a fragment from a gel that you produced in the Restriction Mapping module, hold the mouse down on Restriction Mapping band, then move the mouse to Select band. On the gel that appears click on the band that you want to use then click the selection button at the bottom of the gel.

3. Alternatively, you can select one of the "test" DNA fragments. <u>Click here to view a movie that describes how DNA fragments are isolated from agarose gels.</u>

4. To get another tube, click the rack again and the menu will reappear. You can select as many as 10 DNA fragments. The DNA fragments are at a concentration of 0.2 microgram/microliter.

5. To get rid of a tube, drag it to the trash can underneath the bench.

Task 2: <u>Select the cloning vector.</u>

1. Click the arrow on the vectors rack. Select either <u>pUC19</u>, which is used for making large quantities of DNA, or <u>pMAL</u>, which is used when you want to make protein from the inserted DNA.

2. Select from the submenu both a first and a second restriction enzyme to cut the polylinker, even if they are the same. The restriction enzymes that you use to cut the vector should <u>match</u> the restriction enzymes used to cut the fragment you wish to insert into it. The selected DNA vector is at a concentration of 1 microgram /microliter.

Task 3: <u>Set up the ligation.</u>

1. Click the arrow on the reaction tube rack and select New Empty Tube from the menu. Repeat to get additional tubes for each subcloning reaction.

2. Using the <u>pipettor</u>, transfer the <u>correct amount of vector and fragment</u> to the reaction tube. The correct amount depends on the relative lengths of fragment and vector. In GDL , using 1 microliter of each usually works. Be sure that the restriction sites on the vector and fragment are <u>cohesive</u>.

3. As a control to determine how well the ligation worked, use the vector cut with restriction enzymes but without any insert.

4. Use the pipettor to add the correct amount of <u>buffer and ATP</u> to the reaction tube. For a 20-microliter final volume, add 1 microliter ATP and 2 microliters buffer. The tube will automatically fill with the correct amount of dH_2O.

5. Repeat this process for each ligation reaction you wish to perform.

6. Drag the reaction rack to the right and click on NEXT to move down the bench.

7. Click the arrow on the <u>enzyme cooler</u> and select <u>T4 ligase</u> from the menu.

8. Use the pipettor to transfer 1 microliter of ligase to each reaction tube.

Task 4: <u>Incubate the ligation reaction.</u>

1. Drag the reaction tubes, one at a time, to the <u>water bath</u>.

2. Place the cover onto the water bath, and turn it on by clicking the ON/OFF switch.

3. Click the knob on the water bath and set the temperature for 16 degrees C.

4. Click the time display on the water bath and set the time. When the vector and fragment have cohesive sticky ends, the ligations will generally be completed within 2 hours. When ligating blunt-ended DNAs, it is best to ligate for at least 12 hours.

5. When the timer signals the end of the incubation, drag the reaction tubes back to the rack.

Task 5: Transform bacteria with the ligated DNA.

1. Drag the reaction rack to the right of the screen and click NEXT to move to the next screen. In a cooler are tubes that already contain competent cells.

2. Use the pipettor to transfer 10 microliters of the ligated DNA from each reaction tube to a different tube in the cooler.

3. Set the timer for 15 minutes to allow the ligated DNA molecules to attach to the competent cells.

4. When the incubation is complete, drag the tubes, one by one, to the heat block.

5. Set the temperature on the heat block to 37 degrees C to give a mild heat shock, which promotes entry of the DNA into the bacteria.

6. Set the timer for 5 minutes.

7. After the heat shock is completed, drag the tubes back to the ice bucket.

8. Set the timer for 2 minutes to allow the bacteria to recover from the heat shock. Click here to view an animation on how DNA enters bacteria.

9. When the incubation is completed drag the tubes to the bacteria rack.

Task 6: Plate the transformed bacteria.

1. Drag the bacteria rack to the right and click NEXT to move to the next screen.

2. Use the glass pipette to transfer 3 mls of Luria broth (LB) to each of the tubes. This will allow the bacteria to begin to express the antibiotic resistance gene.

3. Set the timer for 1 hour.

4. On a tray are petri dishes with Luria broth (LB), IPTG, and X-Gal, as well as the antibiotic, ampicillin. Drag each bacterial tube to a different petri dish and release the mouse to dump its contents onto the LB agar.

Task 7: Incubate the transformed bacteria.

1. Click on the door of the bacterial incubator to open it.

2. Drag the tray with the petri dishes on it into the incubator.

3. Click on the door to close it.

4. Set the temperature on the incubator for 37 degrees C.

5. Turn on the incubator by clicking the ON/OFF switch.

6. A new window will open to give you a good view of the petri dishes inside the incubator. As the colonies grow, they will turn white if the DNA fragment has successfully ligated into the vector. If no DNA has been inserted into the vector, it will remain blue.

Task 8: Determine the cloning efficiency.

1. After the bacterial colonies have grown to sufficient size so that the color can be clearly seen, click the film icon to take a picture.

2. Count the number of white colonies and the number of blue colonies on each plate.

3. Record this information in your notebook.

4. If you found no colonies when you expected to find colonies, or if you found only blue colonies on a plate that had been transformed with a ligation reaction in which you had included a DNA fragment to be inserted, consult the troubleshooting section to attempt to identify the reason for the failed subcloning. Remember that if you included, as a control, vector cut with different restriction enzymes and no DNA fragment, then you would expect to get only a few blue colonies.

5. For plates in which you found white colonies, determine the ratios of white to blue.

6. Compare this ratio to the number of colonies you found when you used the vector alone with no fragment. You should see a large increase in the ratio of white to blue colonies when you included fragment, compared to vector alone. If you do see a large increase, it indicates that your subcloning was successful.

7. If the successful subcloning was into pUC19, you can select one of the white colonies in the Sequencing module and determine the sequence of the inserted DNA.

8. If the successful subcloning was into pMAL, you can select a white colony in the Protein Expression module and identify the fusion protein using SDS-polyacrylamide gel electrophoresis (SDS-PAGE).

9. To be sure that you have inserted the DNA fragment in the correct reading frame in pMAL, you should first sequence it in the Sequencing module.

TROUBLESHOOTING

If no colonies appear on a plate, the most likely causes are:
- You forgot to put a cloning vector into the reaction tube.
- You forgot to put a DNA fragment into the reaction tube.
- The cloning vector was cut with restriction enzymes that are different from restriction enzymes used to cut the DNA fragment.
- You forgot to add ligase, ATP, or buffer to the reaction tube.
- The reaction rack was not placed in the water bath.
- The reaction rack was not left in the water bath long enough.
- You forgot to transfer the ligation reaction into the tubes with competent cells.
- You put in the Luria broth before you put in the contents of the reaction tubes .
- You forgot to add the Luria broth.
- If all the colonies on the plate are blue, the most likely causes are:
- The vector was cut with the same restriction enzyme for both cuts and no fragment was added to the reaction tube (the vector will ligate back to itself, at low efficiency, in this case).
- The vector was not cut with any restriction enzymes.

If the ratio of white to blue colonies was high (around 10:1) but the total number of colonies was low (10-30), the most likely reason is:
- The bacterial transformation protocol was not accurately followed.

EQUIPMENT AND REAGENT INFORMATION

Pipettor

Drag the automatic pipettor from its rack. A new disposable tip appears automatically. The tip will be automatically discarded after each use.

Click on the dark rectangular window on the pipettor to get a pop-up slider. Select the volume by adjusting the slider using the sliding needle or by using the arrow buttons for fine control. Click OK or anywhere outside the slider panel to dismiss it.

Drag the pipettor to the rack containing tubes. When the tip of the pipettor is over a tube, the contents of the tube will be visible. This tells you which tube the pipettor tip is in.

When you release the mouse, if the volume is set within the correct limits, the reagent will be drawn up into the pipettor tip. If not, you will get an error message and will need to reset the volume.

Drag the pipettor to the tube to which you want to add the reagent. When the pipettor tip is over the tube, release the mouse. What was in the pipettor tip will now be added to that tube.

If you draw up something you don't want to use, drag the pipettor to the trash. Release the pipettor over the trash and the current tip will be discarded, allowing you to start over with a new tip.

Enzyme Cooler

The enzyme cooler is kept in a freezer at –20 degrees C and maintains this temperature for some time when placed on the bench.

Keeping enzymes cold is important because storage at higher temperatures will cause them to lose their ability to carry out their enzymatic reaction.

Water bath

Turn on the water bath by clicking the ON/OFF switch.

Click the control knob on the water bath to set the temperature. A slider appears that will allow you to set the temperature.

Remove the top of the water bath by dragging it to the bench. After placing the reaction rack in the water bath, drag the top back onto it.

Heat block

Turn on the heat block by clicking the ON/OFF switch.

Click the control knob on the heater and a slider will appear that will allow you to set the temperature.

If you push GO on the timer and you have forgotten to turn on the heat block you will get an error message.

Timer

When you click on the timer a slider appears. Set the slider on the timer for the desired time.

Click the GO button on the timer to start the timer. In GDL, time is speeded up, so the timed process will only take a few seconds.

If you need to adjust anything, click outside the time slider to dismiss it without starting the timer.

Glass pipette

Click on the volume element at the top of the glass pipette to get a pop-up slider.

Select the volume by adjusting the slider using the sliding needle or by using the arrow buttons for fine control. Click OK or anywhere outside the slider panel to dismiss it.

Drag the glass pipette to a bottle or flask. When you release the mouse, if the volume is set within the correct limits, the reagent will be drawn up into the glass pipette. If not, you will get an error message and will need to reset the volume.

Drag the glass pipette to the tube to which you want to add the reagent. When the glass pipette is over the tube, release the mouse. The liquid in the glass pipette will now be added to the tube.

Luria Broth

One liter of Luria broth (LB medium or Luria-Bertani medium) is made by adding to 950 milliliters of water: 10 grams of bacto-tryptone, 5 grams of yeast extract, and 10 grams of NaCl.

The mixture is shaken and the pH is adjusted to 7.0.

To sterilize the mixture it is autoclaved for 20 minutes.

Bacterial Plates

Bacterial plates are made by adding 15 grams of bacto-agar to LB before autoclaving.

After the autoclaved mixture cools sufficiently to be able to hold the flask, any other ingredients, such as IPTG, X-gal, and antibiotics, are added. Cooling is necessary because the added chemicals break down at high temperatures.

The solution is then poured into petri dishes and allowed to harden.

Notebook

This is where you record your observations.

To review any of the photographs you've taken, click the link in the notebook to make the viewing window appear.

You can also use the BACK and NEXT buttons in the viewing window to navigate through your photograph collection.

DNA Sequencing Module

OBJECTIVES

OBJECTIVES

The procedures in this lab will enable you to:

- **Select a DNA fragment to sequence.**
- **Set up the sequencing reaction.**
- **Run a sequencing gel.**
- **Read the DNA sequence.**

PURPOSE

DNA sequencing is used to determine the precise sequence of the nucleotides in a piece of DNA.

You can determine the sequence of the DNA that you subcloned in the Subcloning module or that you amplified in the PCR module. You can use the DNA sequence information to determine the difference between the DNA from normal individuals and the DNA from patients with the diseases under investigation.

To get started immediately, go to <ins>Task 1</ins>.

TABLE OF CONTENTS

Background
Introduction
Conceptual and technical overview
Task 1: Select DNA and primers
Task 2: Add primer and denature the DNA.
Task 3: Allow the primer to anneal to the DNA.
Task 4: Prepare the Label/Enzyme mix.
Task 5: Incubate the sequencing reaction.
Task 6: Stop the sequencing reaction and denature the DNA strands.
Task 7 Load and run the sequencing gel.
Task 8: Determine the sequence of the DNA.
Task 9: Interpreting the sequence.
Troubleshooting
Equipment and reagent information

BACKGROUND

Almost all DNA sequencing performed today uses a method developed by Frederick Sanger in 1977, for which he won the Nobel Prize. Sanger dideoxy sequencing uses the enzyme DNA polymerase and modified nucleotides which terminate the DNA chains made by the DNA polymerase. Using this method, Sanger was the first to determine the sequence of an entire genome, that of a bacteriophage.

APPLICATIONS

One of the most ambitious projects to be attempted in the field of biology is the effort to determine the entire sequence of the human genome. The Human Genome Project encompasses more than this, it also has produced high resolution physical maps of the human genome and a large number of expressed sequence tags (ESTs). ESTs are usually partial sequences from random cDNAs. These sequences can be used to identify genes that are expressed in particular tissues from which the cDNA libraries are made.

DNA sequencing is sometimes the only way to detect certain types of mutations. For example, a point mutation that does not alter or generate a restriction enzyme site, is difficult to detect by any means other than sequencing.

DNA sequencing can reveal the location of signals upstream and downstream of the coding region. For example in the human beta-globin gene, upstream of the coding region the sequence elements for RNA polymerase binding (the TATA box) are found, while downstream there is a polyA addition site.

Task 1, 2, 3, 4, 5, 6, 7, 8, 9

INTRODUCTION

Determination of the precise sequence of the nucleotides in a piece of DNA is important for many different purposes. For example, it allows you to predict the sequence of amino acids in the protein, it provides the basis for determining the defect in a gene responsible for a disease state, it enables you to isolate the gene from other individuals or from other species, and it lets you to produce large quantities of the protein encoded by the gene.

Most DNA sequencing currently performed uses a modified form of DNA polymerase. For DNA polymerase to work requires two things: a piece of single stranded DNA, and a primer that is annealed to the single-stranded DNA. A primer is a short piece of DNA that is complementary to a portion of the DNA that you want to sequence. Because the two strands of DNA are complementary, hydrogen bonds will form between their bases, a process known as annealing. Therefore, at the beginning of this set of experiments, you will need to choose both the piece of DNA that you want to sequence and a primer that is complementary to it.

You can use the DNA sequence information from this module to decide what primers to use in the PCR module, to determine the difference between normal genes and those of patients with certain diseases in the Genomics module or to verify that your expression cloning in the Subcloning module is in the right reading frame.

Task 1, 2, 3, 4, 5, 6, 7, 8, 9

CONCEPTUAL AND TECHNICAL OVERVIEW

Primers

To determine the sequence of a piece of DNA accurately you usually want to sequence both strands. Thus you will want to identify primers that will bind to both of the ends of the piece of DNA. Each of the primers will be used in a separate sequencing reaction.

Be sure that the primers have the correct orientation so that they will bind to the correct DNA strand. To sequence DNA that is subcloned into pUC19, you can use primers that bind just outside the polylinker region.

The primer that is just 5' to the HindIII site in the polylinker is called the pUC19 forward primer. The primer that is just 3' to the EcoRI site in the polylinker is called the pUC19 reverse primer.

Click here to view a detailed map of the pUC19 plasmid

Task 1, 2, 3, 4, 5, 6, 7, 8, 9

Denaturing DNA

The DNA subcloned in a plasmid, or amplified by PCR is double-stranded, and the strands must be separated or "denatured" to allow the primer to bind.

This can be done by either boiling the DNA or adding a chemical that interferes with the formation of the hydrogen bonds that form between the two strands of DNA. In GDL you will use NaOH as a denaturing agent.

Task 1, 2, 3, 4, 5, 6, 7, 8, 9

DNA Annealing

Once the two strands of the DNA are separated, the primer has to be allowed to anneal to it. This is done by allowing the temperature to gradually descend from 80 degreesC to 37degreesC.

This will allow the primer to find the complementary region on the DNA and to anneal to it.

Because the DNA is much longer than the primer, it takes a lot longer for the DNA to find and anneal to its other strand. For this reason, the DNA remains single-stranded except for the short stretch that is annealed to the primer.

Task 1, 2, 3, 4, 5, 6, 7, 8, 9

Sequencing Cocktail

The sequencing reaction consists of the following reagents:
- DNA with its annealed primer
- modified DNA polymerase
- the four nucleotides dATP, dCTP, dTTP and dGTP
- one radioactively labeled nucleotide
- the reducing agent DTT
- buffer

This "cocktail" is then added to the dideoxy terminators.

The labeling mix contains the four normal nucleotides in the correct proportions.

To be able to see the products of the sequencing reaction, they need to be labeled in some way. The most common way to label the DNA when performing nonautomated sequencing is by using a radioactively labeled nucleotide.

We will use ^{35}S-dATP which emits a relatively weak amount of beta radiation that can be detected on X-ray film.

Even though it is a weak radiation source you should be very careful with all radioactive waste.

Dithiothreitol (DTT) is a reducing agent that allows the Sequenase enzyme to work better.

The Sequenase enzyme is a genetic variant of a DNA polymerase encoded in the genome of the T7 bacteriophage. This is the most widely used enzyme for nonautomated sequencing.

Task 1, 2, 3, 4, 5, 6, 7, 8, 9

Dideoxy Sequencing

Enzymatic sequencing works by adding a modified nucleotide, called a dideoxy terminator to the growing chain of DNA produced by the action of the DNA polymerase on the DNA and its annealed primer.

The addition of the dideoxy terminator causes the chain to arrest or terminate. The low ratio of dideoxy terminators to normal nucleotides is what allows the sequencing reaction to work.

Click here to view an animation of the sequencing reaction.

In GDL the dideoxy terminators are already in the wells of the microtiter plate found on screen 4. Each well of the top row contains ddA, each well of the second row from the top contains ddT, each well of the third row from the top contains ddC and each well of the bottom row contains ddG.

Task 1, 2, 3, 4, 5, 6, 7, 8, 9

Polyacrylamide Gel Electrophoresis

Gel electrophoresis is one of the most important tools in molecular biology. It is a means of separating DNA fragments and allows one to calculate their approximate sizes.

DNA has a negative electrical charge provided by the phosphate groups on the DNA backbone. When DNA is in an aqueous solution, hydrogen molecules are removed from the phosphates resulting in a net negative charge.

Because there is one phosphate for each nucleotide, the negative charge is directly proportional to the length of the DNA fragment.

DNA is attracted to the positively charged electrode, with smaller pieces of DNA moving more easily and quickly through the small pores of the gel than larger pieces of DNA.

Very thin polyacrylamide gels permit the separation of DNA molecules that differ by only one nucleotide in length.

The ability to separate DNA molecules that differ by one base is essential for determining DNA sequence as you will see when you read the sequence information.

The detection of one-base differences between different DNA molecules on a polyacrylamide gel is simplified when the DNA is single-stranded.

Moreover, only one strand is labeled with ^{35}S-dATP and that is the strand that was made from the primer that you supplied.

Normally after running the gel you would have to dry it on a gel dryer, place X-ray film over it in a light-tight cassette and expose the film overnight. GDL allows you to see the results directly.

Task 1, 2, 3, 4, 5, 6, 7, 8, 9

PROCEDURES

Task 1: Select DNA and primers.

1. To select a DNA to sequence, click on the arrow near the DNA rack.

2. From the DNA selection menu, select a colony from a bacterial plate that you made in the Subcloning module, or a fragment from a gel run in the PCR module. Alternatively you can select one of the "test" DNAs to sequence.

3. To select another DNA to sequence, click on the arrow to get the menu again and repeat the selection process. Click here to view a movie describing how DNA is prepared from bacterial colonies for sequencing.

4. If you make a mistake you can trash a tube by dragging it to the trash can.

5. To select a primer, click on the arrow on the Primer rack. To sequence a DNA fragment subcloned into the pUC19 plasmid, select either the pUC19 forward or pUC19 reverse primer. Because you should sequence the DNA in both directions, set up one sequencing reaction using the forward primer and a second sequencing reaction using the reverse primer.

6. To sequence a piece of DNA that you amplified in the PCR module, or if your DNA is subcloned in a different plasmid vector select "Create new primer tube".

7. Use the four yellow keys (A, T, C or G) to enter the nucleotides for your primer in the 5' to 3' direction. The primer should be 20 bases long. Remember that the primer sequence must be complementary to the DNA that you wish to sequence and must be in the right orientation.

Task 2: <u>Add primer and denature the DNA</u>.

1. Use the <u>pipettor</u> to add primer to the DNA tube. Add 0.5 microliter (1 picomol) of primer to the DNA. Each DNA tube contains 2.5 microliters of the selected DNA at a concentration of 1 microgram/microliter.

2. Repeat this for each DNA tube. Remember that you want to sequence each DNA in both directions, so add the forward primer to one DNA tube and the reverse primer to another tube with the same DNA in it.

3. The final contents of each tube will be automatically recorded in your <u>notebook</u>.

4. Use the pipettor to add 0.5 microliters of 0.5 N sodium hydroxide (NaOH) to each of the DNA tubes that contain primer. This will <u>denature</u> the DNA.

5. Drag the DNA rack to the <u>incubator</u> underneath the bench. The incubator door will open automatically. Place the DNA rack on the incubator shelf.

6. Click on the knob on the right side of the incubator to set the temperature to 80° C. Turn on the incubator by clicking the ON/OFF switch. High temperature accelerates the denaturation process.

7. Set the <u>timer</u> for 2 minutes.

Task 3: <u>Allow the primer to anneal to the DNA</u>.

1. Turn off the incubator and click on the door to bring the DNA rack back to the bench. Drag the DNA rack to the right of the screen, then click NEXT to move to the next screen.

2. Use the pipettor to add 0.5 microliter of 0.5 N hydrochloric acid (HCl) to each DNA tube. This neutralizes the HCl.

3. Use the pipettor to add 1 microliter of 5x Sequenase buffer (found in the green rack) to each DNA tube.

4. Drag the rack back to the incubator.

5. Set the temperature on the incubator to 37° C and set the timer for 15 minutes. The temperature will gradually descend from 80° C to 37° C. This will allow the primer to find the complementary region on the DNA and to <u>anneal</u> to it.

Task 4: <u>Prepare the Label/Enzyme mix</u>.

1. Turn off the incubator and click on the door to bring the rack back to the bench. Drag the rack to open the tubes.

2. Click on the arrow on the light green rack and select new tube from the menu. To make the "<u>sequencing cocktail</u>" multiply the amount given for each ingredient by the number of DNA samples to arrive at the correct amount.

3. Use the pipettor to add 1 microliter of the <u>labeling mix</u> tube (found in the light green rack) for each DNA that you wish to sequence to the mixing tube. If you are sequencing 4 DNAs, you need to add 4 microliters.

4. Add 0.5 microliter of <u>DTT</u> for each DNA that you wish to sequence to the mixing tube.

5. Add 0.5 microliter of <u>35S-dATP</u> for each DNA that you wish to sequence to the mixing tube. The ^{35}S-dATP is found in a tube behind the radiation protection shield.

6. Click on the arrow on the <u>enzyme cooler</u> and select <u>Sequenase</u> from the menu. Add 1.0 microliter of Sequenase enzyme for each DNA that you wish to sequence to the mixing tube.

Task 5: Incubate the sequencing reaction.

1. Use the pipettor to transfer 3 microliters from the Mixing tube to each of the tubes that contain your DNA annealed to the primer.

2. To give the Sequenase enzyme a chance to find the DNA before it is placed in the presence of the dideoxy nucleotides drag the DNA rack to the incubator.

3. Set the temperature to 37°.

4. Set the timer for 1 minute.

5. When the incubation is complete, turn off the incubator and click on the door to bring the DNA rack back to the bench. Drag the rack to the far right of the screen.

6. Click on NEXT to move the next screen.

7. From one of the tubes that contains the mixture of DNA, primer and labeling cocktail, use the pipettor to transfer 1.8 microliters to the gray 24-well microtiter plate. The wells of the microtiter plate have dideoxy terminators in them. Each well of the top row contains ddA, each well of the second row from the top contains ddT, each well of the third row from the top contains ddC and each well of the bottom row contains ddG

8. Pipette into a well on the side of the plate closest to the top of the screen. The same amount will automatically be placed in each of the wells below it.

9. Repeat this for each of the DNAs that you wish to sequence. Be sure to add each DNA mix to a fresh well across the top row of the microtiter plate.

10. When all the DNA mixes have been added to the microtiter plate, drag the plate to the incubator.

11. Set the incubator for 37 degreesC.

12. Set the timer for 5 minutes. While the sequencing reaction progresses, click here to access an animation that will show you the sequencing reaction that is taking place in the microtiter wells

Task 6: Stop the sequencing reaction and denature the DNA strands.

1. When the incubation is complete, turn off the incubator and click on the door to bring the microtiter plate to the bench.

2. Use the pipettor to transfer 2 microliters of Stop Solution from the tube in the small rack to each of the top wells that have DNA in them.

3. Drag the microtiter plate back to the incubator.

4. Set the temperature for 75 degreesC to denature the double-stranded DNA formed during the sequencing reaction.

5. Set the timer for 2 minutes.

6. Click on the incubator door to bring the microtiter plate back to the bench.

Task 7 Load and run the sequencing gel.

1. Draw up 2 microliters from one of the top microtiter wells and drag it to a well of the sequencing gel. Sequencing gel wells are made with "shark's tooth" combs. The wells are made by the space each "tooth" creates when polyacrylamide is poured to make the gel.

2. When the pipettor tip is over the selected well, release its contents into the well by releasing the mouse. The next three lanes will be automatically loaded with the other wells from the same sequencing reaction. The loading order is A, T, C, G

3. Load all of the sequencing reactions into the gel wells. The contents of each well will be automatically entered into your notebook.

4. When all of the wells are loaded, click on a knob on the power pack. Set the voltage to 2000 volts (V), which should produce about 1000 milliAmperes (mA) of current.

5. Turn on the power supply by clicking the ON/OFF switch.

6. A window will open giving an enlarged view of the gel. You will see the DNA fragments as they would appear on X-ray film. Each labeled fragment of DNA will run on the gel and expose the X-ray film at a specific position, resulting in a dark band on the autoradiogram.

7. While the gel is running, you can adjust the voltage with the slider, and you can stop the electrophoresis by clicking on the STOP button.

8. Take the first picture when the first bands arrive at the bottom of the gel by clicking on the CAMERA button. The picture will be entered into your notebook.

9. Resume the gel running by clicking the RESUME button. Stop and take another picture when the bands that were about two-thirds of the way down reach the bottom of the gel.

10. Repeat the same thing for the third picture.

11. Close the window to return to the bench.

Task 8: Determine the sequence of the DNA.

1. From the picture of the sequence gel you can now "read" the sequence.

2. First, identify four lanes that have the same DNA and primer.

3. Starting at the bottom of the gel, find the lane with a band that is closest to the bottom. In your notebook, type the letter (A,T, C or G) corresponding to this lane. Remember the loading order from left to right is A, T, C, G.

4. Now find the band that is just above the first band. The next band could be in any of the four lanes. For example if the band that is closest to the bottom is in the C lane, the band that is next closest to the bottom could be in the A lane, the T lane, the G lane, or it could be just above the first band in the C lane. By looking across the four lanes you can determine which band is the next. Type the corresponding letter in your notebook immediately after the first letter.

5. Repeat this process of finding the next highest band until you have read as high as possible on that picture.

6. From the next picture of the gel taken when the bands had run further down, continue reading the sequence.

7. As you get to the upper part of the gel of the third picture, the bands will be so tightly packed that it will be difficult, if not impossible to determine which one is higher than another. This is the point at which you can no longer accurately read the sequence.

8. When you have finished the sequence for one DNA, read the sequence of the other DNAs.

Task 9: <u>Interpreting the sequence.</u>

The DNA sequence that you have determined can be used for several purposes:
A: Determining the sequence of a cDNA clone.
B: Determining the sequence of a PCR-amplified piece of DNA.
C: Verifying the reading frame of expression subcloning.

<u>Use A: Determining the sequence of a cDNA clone.</u>

Determining the sequence of a cDNA clone requires two steps: 1) the sequence of each of the subclones must be determined; 2) the sequence of all of the subclones must be aligned to make one continuous sequence. Both steps involve the finding of sequence overlaps, which are regions with the same sequence.

1. You should have sequenced each of your subclones from the cDNA in both directions. To make sure that you have obtained the entire sequence of each of the subclones you must identify a region where the sequence read from each end overlaps.

2. To find overlaps between sequences use the BLAST2Sequences program which you can access from the Genomics module.

3. If you do not find a region of overlap, it means that you have not sequenced far enough into the subclone. Perform two more sequencing reactions, using primers from each end. You will need to run the gel longer this time to be able to read more bases.

4. After finding the overlaps among two sequences, consult the Genomics module to find out how to build up a full-length cDNA sequence.

5. From the sequence of the cDNA you can design primers to be used in the PCR module, or you can use the Genomics module to compare the sequence to other sequences found in public databases.

<u>Use B: Determining the sequence of a PCR-amplified piece of DNA.</u>

1. For PCR-amplified DNA you should follow the same steps as for a cDNA to be sure that you have the complete sequence of the fragment.

2. Once you have determined the entire sequence you can make comparisons between the DNA sequence of the normal individuals with those who exhibit disease symptoms in the Genomics module.

<u>Use C: Verifying subclones.</u>

When subcloning into expression vectors it is essential that the inserted DNA form the correct reading frame with the vector DNA.

1. Because errors can and do occur in DNA ligation and replication be sure to verify the DNA sequence for the correct reading frame before continuing.

2. If the correct primer was used then you should have obtained sequence that covers the point of insertion of the DNA that you wish to have expressed.

3. Compare this sequence to both the vector and the cDNA sequence to be sure that no bases have been added or deleted.

4. Use the ORF Finder program in the Genomics module to translate the DNA sequence into amino acids and make sure that there is the correct open reading frame.

TROUBLESHOOTING:

If no bands appear in a gel lane the most likely causes are:
- ' You forgot to add one of the reagents to the sequencing reaction.
- You forgot one of the incubation steps.
- The primer was not complementary to the DNA in that reaction.

EQUIPMENT AND REAGENT INFORMATION

Pipettor

Drag the automatic pipettor from its rack. A new disposable tip appears automatically. The tip will be automatically discarded after each use.

Click on the dark rectangular window on the pipettor to get a pop-up slider. Select the volume by adjusting the slider using the sliding knob or by using the arrow buttons for fine control. Click OK or anywhere outside the slider panel to dismiss it.

Drag the pipettor to the rack containing tubes. When the tip of the pipettor is over a tube, the contents of the tube will be visible. This tells you which tube the pipettor tip is in.

When you release the mouse, if the volume is set within the correct limits, the reagent will be drawn up into the pipettor tip. If not, you will get an error message and will need to reset the volume.

Drag the pipettor to the tube to which you want to add the reagent. When the pipettor tip is over the tube, release the mouse. What was in the pipettor tip will now be added to that tube.

If you draw up something you don't want to use, drag the pipettor to the trash. Release the pipettor over the trash and the current tip will be discarded, allowing you to start over with a new tip.

Enzyme cooler

The enzyme cooler is kept in a freezer at –20 degrees Celsius and maintains this temperature for some time when placed on the bench.

Keeping enzymes cold is important because storage at higher temperatures will cause them to lose their ability to modify DNA.

Incubator

Turn on the incubator by clicking the ON/OFF switch.

Click the control knob on the incubator to set the temperature. A slider appears that will allow you to set the temperature.

Turn off the incubator and click on the door to remove a rack.

Heat block

Turn on the heat block by clicking the ON/OFF switch.

Click the control knob on the heater to set the temperature. A slider appears that will allow you to set the temperature.

If you push GO on the timer and you have forgotten to turn on the heat block you will get an error message.

Timer

When you click on the timer a slider appears. Set the slider on the timer for the desired digestion time. Click the GO button on the timer to start the timer. In GDL, time is speeded up, so the timed process will only take a few seconds.

If you need to adjust anything, click outside the time slider to dismiss it without starting the timer. If you have forgotten to turn on the piece of equipment being used, a message will appear to remind you.

Stop Solution

Addition of the Stop solution will stop the sequencing reaction.

It also supplies a blue dye to visualize the progress of the electrophoresis in the gel. Because in GDL you can see the DNA directly, you don't need to use running dye for this purpose.

Notebook

This is where you record your observations.

To review any of the gel photographs you've taken, click any gel/lane number (such as "A02") in the records in the notebook to make the viewing window appear.

You can also use the Back and Next buttons in the viewing window to navigate through your photograph collection.

PCR Module

<u>OBJECTIVES</u>

The procedures in this lab will enable you to:

- **Select DNA and compose primers for PCR amplification.**
- **Use PCR to amplify DNA samples.**
- **Separate amplified DNA fragments by gel electrophoresis.**

<u>PURPOSE</u>

PCR is used to amplify DNA using primers that are complementary to the DNA.

You can design primers based on the sequence generated in the DNA Sequencing module, and use them to amplify from DNAs isolated from individuals with the diseases under investigation.

To get started immediately, go to Task 1.

TABLE OF CONTENTS

<u>Background</u>
<u>Introduction</u>
<u>Conceptual and technical overview</u>
<u>Task 1: Select DNA and Primers for PCR amplification.</u>
<u>Task 2: Prepare the Primer/Polymerase mix.</u>
<u>Task 3: Add DNA and Reaction mix to PCR tubes.</u>
<u>Task 4: Start the PCR reaction.</u>
<u>Task 5: Add running dye and load the gel.</u>
<u>Task 6: Separate the PCR amplified DNA on an Agarose gel.</u>
<u>Task 7: Interpret the results.</u>
<u>Troubleshooting</u>
<u>Equipment and reagent information</u>

BACKGROUND

The polymerase chain reaction (PCR) was introduced in 1987 by Kary Mullis, who later won a Nobel Prize for this technique which has revolutionized biology. It is similar to cloning in the sense that it is a method that makes many copies of a DNA molecule. Unlike cloning, however, which requires ligating pieces of DNA together and then putting the recombinant DNA into bacteria to performs the amplification, PCR is done in a test tube in a PCR machine or thermal cycler. It is therefore much faster and easier than cloning.

<u>APPLICATIONS</u>

PCR has many applications in genetics and forensics. Because of its speed and ease of use PCR is frequently utilized in genetic testing. For example, disease gene alleles can be tracked using PCR amplification. Inheritance of the dominant allele for Huntington's disease was analyzed using PCR to detect a trinucleotide expansion responsible for the disease state. Another application of PCR utilizes single short primers to generate DNA markers in a method called random amplified polymorphic DNA (RAPD) analysis.

Because PCR is able to amplify from very small quantities of DNA, it has been used to analyze ancient DNA, for example DNA isolated from ancient insects preserved in amber. PCR has also been used to amplify DNA from Egyptian mummies. From the comparison of the DNA sequences it has been possible to reconstruct the lineage of the Egyptian royal family.

Task <u>1</u>, <u>2</u>, <u>3</u>, <u>4</u>, <u>5</u>, <u>6</u>, 7

INTRODUCTION

The technique known as polymerase chain reaction (PCR) has revolutionized biology. Using PCR, a researcher can amplify a piece of DNA in a few hours. Once amplified, the DNA can be sequenced, subcloned, restriction mapped, etc.

You will be able to choose as targets for PCR amplification the genomic DNA from individuals with sickle-cell anemia, beta-thalassemia, and certain cancers, as well as DNA from healthy individuals. You will separate the products of PCR amplification using <u>agarose gel electrophoresis</u>. In some cases, the differences between individuals with a disease and their healthy counterparts may be apparent from differences in the length of the amplified fragments. In other cases, differences may only be detectable when the amplified DNA is sequenced or digested with restriction enzymes. You can perform these experiments in the DNA Sequencing module or in the Restriction Mapping module.

Task <u>1</u>, <u>2</u>, <u>3</u>, <u>4</u>, <u>5</u>, <u>6</u>, <u>7</u>

CONCEPTUAL AND TECHNICAL OVERVIEW

PCR Amplification

PCR amplification is based on how DNA polymerase works – it acts on template DNA that has a short primer, or oligonucleotide, that is annealed to it through complementary base-pairing (A to T, and C to G). DNA polymerase will recognize the primer bound to DNA and extend the primer, adding nucleotides until it reaches the end of the piece of DNA.

In PCR, when DNA polymerase has copied one strand of DNA, the now double-stranded DNA is heated to a temperature at which the two strands come apart or denature.

This is a little hard to follow because written "backwards." A clearer version might be: The first primer used, called the forward primer, binds to the original strand. Another primer, known as the reverse primer, binds to the newly synthesized strand.

Now there are two strands and two primers, with each primer going in the opposite direction. <u>Click here for an animation of the PCR process.</u>

Once polymerase acts on these DNAs, there will be two double-stranded molecules. These are then denatured, and the process of extending from the two primers starts again.

Since two DNA molecules are produced from one in each PCR cycle, this is an exponential process – you start with 1 DNA strand, then you get 2, then 4, then 8, then 16, etc. This is why you can make a lot of DNA in a short time. It is also why you can use PCR to detect minute quantities of DNA.

Task <u>1</u>, <u>2</u>, <u>3</u>, <u>4</u>, <u>5</u>, <u>6</u>, <u>7</u>

Primer Design Guidelines

To amplify a particular piece of DNA, you will need to design forward and reverse primers that can bind to the 5′ and 3′ ends of the template DNA. The same applies if you want to amplify a region of genomic DNA: You need to choose primers that flank that region.

To design PCR primers that are specific for a particular DNA, you need to know the sequence of the DNA. Remember that the forward primer will bind to one strand and the reverse primer will bind to the other strand of the DNA you wish to amplify.

PCR primers need to be long enough so they will bind only to the target DNA and not to other DNAs. There is a greater chance that a longer sequence is found only once in the target DNA. Usually a primer of 20 nucleotides is long enough to ensure that it will bind specifically to the target DNA.

Another factor in determining the specificity of binding is the number of G and C nucleotides in the primer, compared to the number of A and T nucleotides. This is known as the GC content of the primer.

The higher the GC content, the more tightly the primer will bind to the DNA and the more specific it will be. This is because the interaction between A and T nucleotides is mediated by only two hydrogen bonds, while the interaction between G and C nucleotides is mediated by three hydrogen bonds, making the interaction harder to disrupt.

For best results, when designing your primer, look for regions of your target DNA that have approximately 50% GC content.

Task 1, 2, 3, 4, 5, 6, 7

PCR Cycles and Thermostable Polymerase

One PCR cycle consists of the following three steps [or stages]:
- Primers anneal to target DNA.
- Polymerase extends the DNA strands.
- Denaturation of the newly formed double-stranded DNA.

The number of cycles required to amplify enough DNA to make it visible on an agarose gel depends primarily on the amount of target DNA in the initial reaction. The more target DNA you start with, the fewer cycles you will need to make sufficient amounts to see on a gel.

Because the DNA has to be denatured with high temperature at every cycle, there are two choices for DNA polymerases. Most DNA polymerases will be destroyed by high temperature because the protein unfolds. In the original PCR methods, fresh polymerase had to be added at each cycle.

Then it was discovered that bacteria that live in high-temperature environments, such as geysers, have polymerases that can withstand very high temperatures without unfolding. Today, all PCR methods use these "thermostable" polymerases.

Task 1, 2, 3, 4, 5, 6, 7

Temperature of PCR Amplification

For PCR to amplify DNA, the temperature of the reaction must be changed for each part of the cycle.

The first step of annealing requires a temperature that is low enough for the primer to anneal to the target DNA, but high enough to prevent it from annealing to sequences that are similar to but not exactly the same as the target.

The optimal temperature depends on the GC content of the primer. A guideline for calculating the annealing temperature is degrees Celsius = 4 x (G + C) + 2 x (A + T). This means you count the number of G and C nucleotides in the primer and multiply that number by 4, then count the number of A and T nucleotides and multiply that number by 2, then add these two numbers together.

For example, if there are 10 Gs and Cs and 10 As and Ts, then 4 x 10 + 2 x 10 = 60 degrees C as the optimum annealing temperature for this oligonucleotide.

Task 1, 2, 3, 4, 5, 6, 7

PCR Cycle Times

Annealing of primers to target DNA usually only takes 30 seconds. Then the temperature must be raised for the polymerase to add nucleotides, which is called the elongation step.

Elongation is usually performed at 72 degrees C. The length of time that elongation takes depends on the length of the DNA you wish to amplify.

For up to 1000 base-pairs (1 kilobase, or 1 kb), 40 seconds is generally enough to allow for complete elongation. For fragments longer than 1 kb, elongation times must be increased proportionally.

Denaturing the DNA strands is performed at 94 degrees C. This usually takes only 20 seconds.

Task 1, 2, 3, 4, 5, 6, 7

Agarose Gel Electrophoresis

Gel electrophoresis is a means of separating DNA fragments. It enables one to calculate their approximate sizes.

DNA has a negative electrical charge provided by the phosphate groups on the DNA backbone. When DNA is in an aqueous solution, hydrogen molecules are removed from the phosphates, resulting in a net negative charge.

Because there is one phosphate for each nucleotide, the negative charge is directly proportional to the length of the DNA fragment.

Agarose gel electrophoresis is performed in a gel box into which the gel is placed. A current is run through a buffer solution surrounding the gel, with the negative electrode near where the DNA is added to the gel, and the positive electrode at the other end of the gel box.

DNA is attracted to the positively charged electrode, with smaller pieces of DNA moving more easily and quickly through the pores of the gel than larger pieces.

Agarose gels are typically used to separate PCR-amplified fragments that are larger than 200 bp.

A gel of the size used in this module normally runs at about 100 volts (V), which produces about 120 milliAmperes (mA) of current.

Task 1, 2, 3, 4, 5, 6, 7

Determining DNA Fragment Sizes

To get an estimate of the size of a PCR-amplified piece of DNA, compare its position on the gel with the position of the gel marker DNA bands. Those that appear at about the same place have about the same size.

To accurately determine the size of a DNA fragment on a gel, first use a ruler to measure the distance from the starting well to each marker DNA fragment, and plot this information on a piece of graph paper or in a graphing program.

Make the Y axis the logarithm of fragment size and the X axis the distance migrated. Draw a straight line through the plotted points.

Determine the distance each amplified fragment has migrated in the gel.
Plot each of these distances on the graph marker DNA size vs. distance, and determine the approximate size of the fragment.

It is also very common to directly compare DNA fragments with marker DNA bands that are running at a similar level on the gel. This usually gives a good estimate of fragment size.

Task 1, 2, 3, 4, 5, 6, 7

PROCEDURES

Task 1: Select DNA and primers for PCR amplification.

1. To select a target DNA for amplification, click the arrow on the DNA rack. To determine the genetic defect in individuals with a particular disease, you will want to compare their DNA to that from a healthy counterpart.

2. On the menu, select DNA from the beta-globin or p53 genomic locus of a normal individual and from patients with the diseases you wish to study.

3. To get another tube, click the arrow again and the menu will reappear. You can select as many as 10 DNA tubes. The concentration of the genomic DNA is 50 nanograms/microliter. Click here to find out how genomic DNA is extracted.

4. If you have made a mistake, drag the tube to the trash can.

5. To select a primer, click the arrow on the Primer rack. To design good primers , you must know at least some of the sequence of the target DNA and take into account the GC content. You should use the sequence you determined in the DNA Sequencing module to design the primers.

6. Use the four yellow keys (A, T, C, or G) to enter the nucleotides for your primer in the 5′ to 3′ direction. The primer should be 20 bases long. Remember that the primer sequence must be complementary to the DNA you wish to amplify, and it must be in the right direction.

7. To make another primer, click on the Primer rack to get the menu again. Remember that you will need both a forward and a reverse primer for each amplification reaction. Primers are at a concentration of 0.5 microgram/microliter.

Task 2: Prepare the primer/polymerase mix.

1. Click on the arrow on the mixing tube rack and select New Empty Tube from the menu. You will need a mixing tube for each set of primers you will be using.

2. To make the PCR cocktail, multiply the amount given for each ingredient by the number of DNA samples to arrive at the correct amount. Use the pipettor to add 2 microliters of the dNTPs to the mixing tube for each DNA you will amplify. The tube of dNTPs is in the reagent rack and consists of a stock of 10 mM of dATP, dCTP, dTTP, and dGTP.

3. Use the pipettor to add 10 microliters of 10x amplification buffer (in the reagent rack) to the mixing tube for each DNA.

4. Add 85.7 microliters of dH_2O (in the reagent rack) to the mixing tube for each DNA. Do this sequentially, because the pipettor's capacity is only 120 microliters.

5. Click the arrow on the enzyme cooler. On the menu, select thermostable polymerase.

6. Add 0.3 microliter of thermostable polymerase to the mixing tube for each DNA.

7. Add 1 microliter of forward primer to the mixing tube for each DNA.

8. Add 1 microliter of reverse primer for each DNA you wish to sequence to the mixing tube.

Task 3: Add DNA and reaction mix to PCR tubes.

1. Press NEXT to move to the next screen.

2. Click the arrow on the PCR rack to select PCR tubes. You will need a PCR tube for each DNA you wish to amplify.

3. Use the pipettor to add 1 microliter of DNA to the first PCR tube.

4. Add 100 microliters of the reaction mix that contains the primers you wish to use to amplify the DNA in the first PCR tube.

5. Repeat this for each DNA tube. You can add the same reaction mix to more than one DNA tube. However, you cannot add two different reaction mixes to one DNA tube.

6. Because PCR is able to amplify minute quantities of DNA, a few molecules of stray DNA that find their way into a reaction tube can give false results. To determine if there is contaminating DNA, for each set of primers, add the reaction mix to a PCR tube with no DNA.

Task 4: Start the PCR reaction.

1. Drag the PCR tubes one by one into the rack in the PCR machine.

2. Click on the cover to close it.

3. Click on the front panel of the PCR machine to get the set-up menu.

4. Set each of the parameters by releasing the mouse when the parameter is highlighted on the menu. Use the slider that appears to set the temperature, time, or number of cycles.

5. For the annealing temperature, choose the temperature that corresponds to the lower temperature calculated for each of your two primers. Set the time for the annealing step to 30 seconds.

6. The elongation temperature is normally set to 72 degrees C. The time depends on the length of the DNA to be amplified. For 1 kb or less, 40 seconds is enough.

7. The denaturation temperature is normally set to 94 degrees C. The time for denaturation is 20 seconds.

8. Set the number of cycles. For PCR amplification from genomic DNA, 30 is usually sufficient.

9. When all the steps have been programmed, select Start PCR from the menu.

10. While the DNA is amplifying, you can access an animation that will show you what's going on in the tubes.

Task 5: Add running dye and load the gel.

1. When the amplification reaction is finished, click NEXT to move to the next screen. Click the door of the PCR machine to open it, and drag the PCR tubes to the rack on the bench.

2. Use the pipettor to add running dye or blue dye (in the small rack) to each of the PCR tubes. The blue dye is in a 10x concentrated form, so for a 100-microliter reaction, add 10 microliters.

3. Click NEXT. You will now use agarose gel electrophoresis to separate the DNA fragments. Click here to view a movie on how to prepare an agarose gel.

4. Use the pipettor to transfer 10 microliters of gel marker (in the small rack) to the first gel well on the left. When the pipettor tip is over the well, release its contents by releasing the mouse button to at least one of the gel wells.

5. Use the pipettor to transfer 15 microliters of the amplified DNA (to which you have added blue dye) to a different well of the gel.

6. The contents of each well are revealed when the pipettor rolls over them, and they will be automatically entered into your notebook.

Task 6: Separate the PCR-amplified DNA on an agarose gel.

1. When all the wells are loaded, click NEXT.

2. Drag the cover onto the gel box. It is correctly positioned when you hear it click into place.

3. Adjust the voltage by clicking on the voltage/amperage indicator. 100 volts is an appropriate voltage for this gel.

4. Turn on the power supply by clicking the ON/OFF switch. A new window will open, giving you a good view of the gel inside the gel box.

5. After a few moments, an ultraviolet light will illuminate the gel in which ethidium bromide binds to the DNA fragments and glows orange.

6. If necessary, you can adjust the voltage while the gel is running.

7. You can stop and resume the electrophoresis at any time. The goal is to get maximum separation of the fragments without letting any of them run off the bottom of the gel. Fragments that do run off the end of the gel are lost.

8. Click on the camera icon to take a picture of the gel. The picture will be entered into your notebook. You can take up to three pictures of each gel.

9. Close the window to return to the bench.

Task 7: Interpret the results.

The results of PCR amplification can be interpreted using several different techniques. Usually the first thing that is done after amplification is to run an agarose gel to make sure that amplification has actually occurred.

Comparison of the amplification products separated on an agarose gel can provide a lot of information. First, you want to compare the size of the fragment you generated from genomic DNA with the expected size based on the cDNA.

1. To determine the size of the amplified DNA fragment , you can estimate it by comparing its location on the gel to the gel marker fragments, or you can graph the marker fragments locations on semi-log paper.

2. If the size of the amplified product is larger than that of the size expected from the cDNA, it could mean that there is an intron in the genomic DNA between the two primers. To determine whether this is the case, you can sequence the amplified DNA in the DNA Sequencing module.

3. Comparing the amplified DNAs to each other can be useful for determining the genetic defect responsible for a disease. Compare the size of the DNA fragment from the healthy individual with the size of the DNA fragments from the individuals with disease that were amplified using the same primers. If there are differences, this could indicate that a deletion occurred in the DNA of the individuals with a disease.

4. To verify this, sequence the DNA in the DNA Sequencing module.

5. If there are no apparent differences, you will need to sequence the DNA from the individuals with a disease, or try to find a restriction enzyme that cuts in one DNA but not the other. Sequencing is usually the faster approach to find differences in the DNA. By comparing the DNA sequence of the healthy individual with the DNA sequence of the individuals with a disease, you may discover differences in the DNA sequence.

6. Once you have found a good candidate for a disease –causing mutation, you can attempt to devise a rapid diagnostic method for detecting that mutation. If the change in the DNA sequence causes a restriction enzyme site to be deleted or created, you can use restriction enzyme analysis as a rapid diagnostic for that mutation.

7. This is done by PCR amplifying the DNA from healthy and disease-carrying individuals. After the DNA is amplified, it is cut with restriction enzymes and run out on agarose gel. Differences in the pattern of the fragments on the gel can be used to rapidly identify individuals with the mutation.

TROUBLESHOOTING

If a band from amplified DNA does not appear in a gel lane, the most likely causes are:
- Either the forward or the reverse primer was left out or was not complementary to the DNA in that reaction.
- One of the reagents was not added to the reaction.
- The annealing temperature was too high for the GC content. Remember the annealing temperature is calculated as $4 \times (G + C) + 2 \times (A + T)$.

EQUIPMENT AND REAGENT INFORMATION

Pipettor

Drag the automatic pipettor from its rack. A new disposable tip appears automatically. The tip will be automatically discarded after each use.

Click on the dark rectangular window on the pipettor to get a pop-up slider. Select the volume by adjusting the slider using the sliding needle or by using the arrow buttons for fine control. Click OK or anywhere outside the slider panel to dismiss it.

Drag the pipettor to the rack containing tubes. When the tip of the pipettor is over a tube, the contents of the tube will be visible. This tells you which tube the pipettor tip is in.

When you release the mouse, if the volume is set within the correct limits, the reagent will be drawn up into the pipettor tip. If not, you will get an error message and will need to reset the volume.

Drag the pipettor to the tube to which you want to add the reagent. When the pipettor tip is over the tube, release the mouse. What was in the pipettor tip will now be added to that tube.

PCR Machine

Click on the PCR machine door to open or close it.

To set each of the parameters, click on the PCR machine display area to get a menu. Set each of the parameters by releasing the mouse when the desired parameter is selected: annealing temperature, annealing time, elongation temperature, elongation time, denaturing temperature, denaturing time, number of cycles.

You will get a slider that you can set for the desired time or temperature. When all the parameters are set, turn on the PCR machine using the menu.
The PCR machine display will indicate the cycle that is under way.

Enzyme Cooler

The enzyme cooler is kept in a freezer at –20 degrees C and maintains this temperature for some time when placed on the bench.

Keeping enzymes cold is important because storage at higher temperatures will cause them to lose their ability to enzymatic activity.

Gel Marker

The gel marker is DNA from bacteriophage lambda that has been cut with the restriction enzyme HindIII.

This marker has been used for many years because the resulting fragments produce a nice "ladder" of known sizes.

The sizes of the marker DNA from largest to smallest are: 23,130 base-pairs (bp), 9416 bp, 6557 bp, 4361 bp, 2322 bp, 2027 bp, 564 bp, and 125 bp.

Running Dye

The running dye or "blue dye" contains dextran sulfate, a substance that makes the DNA sink to the bottom of the wells in the gel.

If you didn't add it, the DNA would come out of the well and disappear in the buffer.

Blue dye also allows you to see how fast the DNA is moving through the gel. Because you can visualize the DNA directly in GDL, you don't need to use running dye for this purpose.

Notebook

This is where you record your observations.

To review any of the gel photographs you've taken, click any gel/lane number (such as A02) in the records in the notebook to make the viewing window appear.

You can also use the BACK and NEXT buttons in the viewing window to navigate through your photograph collection.

Genomics Module

The procedures in this lab will enable you to:

- **Use Web resources to align DNA sequences.**
- **Find open reading frames (ORFs) in a DNA sequence.**
- **Perform a BLAST search of GenBank.**

PURPOSE

Genomics techniques are used to analyze and compare DNA sequences.

You can use resources available on the Internet to analyze the DNA sequences you generated in the Sequencing module.

To get started immediately, go to <u>Task 1</u>.

TABLE OF CONTENTS

<u>Background</u>
<u>Introduction</u>
<u>Conceptual and technical overview</u>
<u>Task 1: Align overlapping DNA sequences</u>
<u>Task 2: Build a full-length DNA sequence</u>
<u>Task 3: Find open reading frames (ORFs)</u>
<u>Task 4: Interpret the ORFfinder results</u>
<u>Task 5: Perform a BLAST search on a DNA sequence</u>
<u>Task 6: Read the BLAST search results</u>
<u>Task 7: Interpret the BLAST search results</u>
<u>Task 8: Explore genomic resources on the Web</u>

BACKGROUND

Determination of the complete nucleotide sequence of an organism is the basis for the field of genomics. Knowledge of the complete genomic sequence enables scientists to predict all of the genes that are required for the functioning of the organism. In addition to the polypeptide coding regions, genomic sequence permits the analysis of potential regulatory regions upstream and downstream of coding regions. The first genome sequence, that of a bacteriophage was published in 1977. It was not until 1995 that the first genomic sequence of a bacterium was published. The first eukaryotic genome, that of baker's yeast, *Saccharomyces cerevisiae* was completed in 1996. Recently, the sequence of the roundworm, *C. elegans*, the fruit fly, *Drosophila melanogaster* and the plant model system *Arabidopsis thaliana* have all become available.

Advances in sequencing, particularly the use of automated sequencing have been critical to the ability to acquire the DNA sequence of large eukaryotic genomes. In automated sequencing, each dideoxy terminator is labeled with a different fluorescent dye. As the labeled strands of DNA are separated by gel electrophoresis or through a capillary, a sensor detects each labeled strand as it passes a particular point in the gel or in the capillary.

APPLICATIONS

One of the first goals of the Human Genome Project was to produce a high-density physical map of the human genome. This has accelerated the discovery of disease genes such as the gene responsible for cystic fibrosis. Starting from physical landmarks on the genome the cystic fibrosis gene was found through a combination of chromosome walking and jumping [pp. 80-83]. As more such genes are discovered, effective treatments for genetic disorders can be developed.

From the complete sequence of an organism, scientists can learn about whole genome organization. For example, in baker's yeast, it was found that there are regions in which a particular arrangement of genes is repeated in several places in the genome. This has been called a cluster homology region (CHR). The task of identifying the function of every gene in the genome is the next major challenge of genomics.

Task 1, 2, 3, 4, 5, 6, 7, 8

INTRODUCTION

The programs you will use in this module are found on the Web site maintained by the National Center for Biotechnology Information (NCBI), which is a branch of the National Institutes of Health (NIH). The NCBI is responsible for an extensive database of DNA and protein sequence known as GenBank. As researchers discover new sequences in any organism, they deposit these sequences into this public database. The NCBI also develops and makes publicly available many tools for DNA sequence analysis. You will use some of these tools to analyze your DNA sequences.

The NCBI Web site also has a lot of information related to genomics. The term genomics refers to efforts to determine the entire sequence of DNA in an organism and to determine the function of all of its genes. One of the most visible genomics projects has as its goal the determination of the complete sequence of the human genome: all the DNA in all the human chromosomes. This is known as the Human Genome Project.

You can also explore the genomics resources available on the NCBI Web site. For example, the site contains information about the chromosomal position of genes that are implicated in various diseases. You can follow links of some of these genes, including p53, to articles about them that can be accessed through the free bibliographic search engine, PubMed.

Task 1, 2, 3, 4, 5, 6, 7, 8

CONCEPTUAL AND TECHNICAL OVERVIEW

Genomics

The field of genomics began with the goal of determining the complete sequence of the human genome. As a first step toward this goal, the Human Genome Project, a consortium of publicly funded research groups, completed a physical map of all of the human chromosomes. This effort also greatly helped in the identification of disease genes.

The next step in sequencing the human genome was to break up the DNA into manageable pieces. Pieces of DNA ranging from 100 kilobases (kb) to more than a megabase (1000 kb) were subcloned into vectors, such as cosmids or yeast artificial chromosomes (YACs). Using restriction mapping techniques, the overlaps between the subclones were identified. Knowledge of the overlaps enabled scientists to line up the pieces of DNA in what are called contigs. Click here to view an animation that illustrates how one goes from making a physical map to forming contigs of DNA.

The DNA in a cosmid or YAC was then broken into even smaller pieces by random shearing of the DNA, and these pieces were subcloned into plasmid vectors. This DNA was sequenced using automated DNA sequencers. These sequencing machines use essentially the same process you used to do manual sequencing in the Sequencing module. But instead of making the DNA radioactively labeled, a fluorescent tag is used for each of the terminating nucleotides. The tagged DNA is separated using either a polyacrylamide gel or a capillary tube, and each of the four fluorescently labeled bases is detected as it passes a particular point on the gel or capillary. Click here to view a movie about automated sequencing.

Methods now exist that can sequence the human genome and other genomes by skipping the time-consuming steps of producing a physical map and making contigs based on the map. Instead, all the DNA in the genome is randomly sheared and subcloned. By sequencing many pieces of DNA that cover the same region, overlaps can be determined, and a full-length sequence from one tip of a chromosome to the other can be assembled.

With the mapping of the human genome nearly complete, and that of several other smaller genomes, attention is now turning to using this information to determine the function of genes. This subfield is referred to as functional genomics. One of the first steps in determining gene function is understanding when and where genes are expressed in an organism. The information from the already completed genomes is being used to produce arrays of DNA corresponding to all known genes on glass slides. These "microarrays" are then hybridized with labeled RNA in a process similar to Northern blotting, and the level of RNA expression is thus determined for every gene in the genome.

Task 1, 2, 3, 4, 5, 6, 7, 8

Bioinformatics

The field of genomics generates massive amounts of data. The data are in the form of both DNA sequence information and readouts from microarrays and other functional genomics experiments. Analysis of the data requires the expertise of scientists trained in a more mathematical or computer science approach. This has resulted in the development of another new field known as bioinformatics.

Task 1, 2, 3, 4, 5, 6, 7, 8

Alignment of Two Sequences: BLAST 2 Sequences

One of the most widely used bioinformatics programs is called BLAST. It uses a rapid and efficient strategy to find similar DNA or protein sequences.

The BLAST 2 sequences program finds the overlap between two DNA sequences by comparing the two sequences until if finds similar (or identical) sequences. It presents the results in both graphical and text formats.

Task 1, 2, 3, 4, 5, 6, 7, 8

Open Reading Frames

The regions of sequence that are likely to code for a protein can be determined by locating open reading frames (ORFs) in the DNA sequence. You will use a program called ORF Finder to identify ORFs in a DNA sequence. The program translates each codon (a set of three bases in the ORF) into the appropriate amino acid. It uses the standard genetic code, unless a different code is specified.

The program continues to translate three-base codons into amino acids until it reaches a stop codon. In ORF Finder, you can set the minimum number of bases that the program considers to be a real ORF. The default is 100. In this case, if there are fewer than 100 bases (33 amino acids) before it finds a stop codon, then the putative ORF will not be displayed. You can change the minimum number to 50 or 200.

In a double-stranded piece of DNA, ORFs can exist in any one of 6 possible reading frames. The reading frame that starts with the first base in the strand of DNA you entered into the program is called +1. The reading frame that starts with the second base in this direction is called +2, the one that starts with the third base is called +3. The reading frames found on the opposite strand of DNA are given negative numbers (-1,-2,-3).

Several other more complex programs look for ORFs, intron splice sites, poly-A addition sites, and other signs that a gene resides in a stretch of genomic DNA sequence. Because about 40% of all genes have not yet been characterized, these gene finder programs are one of the primary ways of identifying putative genes in genomic sequence.

Task 1, 2, 3, 4, 5, 6, 7, 8

Finding Related Genes: BLAST Searches of GenBank

The BLAST program is used extensively to compare DNA or protein sequences to sequences found in databases such as GenBank. A good way to learn more about the BLAST program is to follow the tutorial found at: http://www.ncbi.nlm.nih.gov:80/BLAST/tutorial/Altschul-1.html/

An important scientific question arises when you find your sequence is similar to other sequences in GenBank: Are the two sequences related by descent? If two sequences are similar, it often means they were derived from the same ancestral sequence during evolution. It also can mean that they encode proteins with similar function.

To determine if sequences are similar or homologous, the BLAST program looks for short stretches of sequence identity. It then looks to see how far in both directions the identity extends. The program will introduce spaces or gaps if that will allow it to extend the region of identity. The program will also allow a certain number of mismatches between the sequences as it extends in either direction. The size and number of gaps and mismatches that can be tolerated are parameters, which can be set in the program.

Depending on the nature of the starting sequence and the type of search to be conducted, different BLAST programs are used. For example, the blastn program compares a DNA sequence to DNA sequences in the database, while the blastp program compares a protein sequence to protein sequences in the database. The blastx program will first translate a DNA sequence into the correct amino acid sequence for all reading frames, then compare that to the protein databases. The tblastn program compares a protein sequence to a DNA sequence database that has been translated into all 6 reading frames.

Each similar sequence that BLAST finds is assigned a probability, or a likelihood of occurring by pure chance. The results of a BLAST search are given in the order of lowest to highest probability. In other words, the top "hit" on the list from a BLAST search is the one most likely to represent a truly related sequence. The extent to which the similarity is likely to be just a random line-up is represented by the E-value. The smaller the E-value, the less likely it is random, and the more likely it is a meaningful similarity.

PROCEDURES   Task 1, 2, 3, 4, 5, 6, 7, 8

Task 1: Align overlapping DNA sequences.

1. Open your Web browser to the NCBI Web site on the BLAST page
    http://www.ncbi.nlm.nih.gov:80/BLAST/.

2. In the Sequencing module, you determined the DNA sequence of several overlapping restriction fragments. To find where these fragments overlap, click on "BLAST 2 sequences."

3. Open your Sequencing module notebook and copy one of the sequences you wish to align. Paste it into the upper open box.

4. Copy and paste the second sequence from your notebook that should overlap the first one into the lower box.

5. Click the ALIGN button at the bottom of the Web page. You will get a graphic representation of the overlap (if any) between the two sequences. At the bottom of the page is the DNA sequence that overlaps.

Task 2: Build a full-length DNA sequence.

1. Build a full-length DNA sequence from fragments by lining up the fragments end to end using regions of overlap.

2. Using the information on where two fragments overlap, eliminate the overlapping sequence from one of the two fragments, and then join the remaining sequence together.

3. Do this for two overlapping sequences in your notebook. Then copy this sequence into one of the "BLAST 2 sequences" boxes. Copy from your notebook another sequence from a fragment that overlaps either one of the ends of this sequence into the other box, and perform the alignment.

4. Repeat the process of finding and eliminating overlaps until you have joined together all the fragments that came from the same original clone.

5. Once you have assembled the DNA sequence for the same gene from a healthy individual and from someone with one of the diseases you are studying, you can use the BLAST 2 sequences tool to compare the two sequences.

Task 3: Find open reading frames (ORFs).

1. Open your notebook and find the DNA sequence you wish to analyze. This could be the full-length sequence you just built by joining overlapping fragments.

2. To find the open reading frames in this sequence, go to the NCBI Web site on the Tools page: http://www.ncbi.nlm.nih.gov:80/Tools/

3. Click on "ORF Finder" on the Tools page.

4. Copy the DNA sequence from your notebook and paste it into the box that says "Sequence in FASTA format."

5. Click on "ORF Find." You will get a graphical depiction of the ORFs found in the sequence you entered, as well as a list of the ORFs with their lengths.

6. The settings above the rectangles allow you to choose the minimum length of an ORF. Change from 100 (the default) to 50, and then click on Redraw to find all ORFs that are at least 50 nucleotides in length.

7. To see the amino acid sequence found in the ORF, click on one of the green boxes that represents the ORF within the total sequence (represented by the black outlined box). The amino acid sequence (in one letter code) will appear beneath the nucleotide sequence.

Task 4: Interpret the ORFFinder results.

1. Select one of the open reading frames you think most likely represents the protein sequence encoded by the DNA. In some cases, this is a simple choice. For example, if the DNA sequence is from a cDNA, there should be no interruptions in the ORF. Usually, but not always, the correct ORF from a cDNA is the longest uninterrupted ORF that begins with a methionine.

2. Once you have found what you think is the correct ORF, highlight it in on the Web page, and copy and paste it into your notebook.

3. If you copy and paste it into a word processing program, you will probably want to use the Courier font to maintain the correct spacing between the nucleotide sequence and the amino acid sequence. Even in this format, you may have to add some spaces so that the alignment is accurate. The amino acid code should be just below the first letter of its codon.

4. Once you have decided which ORF is likely the correct one, click ACCEPT, below the graphical representation of the ORFs.

5. On the next Web page, click VIEW just above the graphical representation of the ORFs. This will give you a text version of the ORF, which you can copy to your notebook or print.

6. Compare the ORF you have chosen with the amino acid sequence of p53 or beta-globin found in the cDNA Cloning module.

Task 5: Perform a BLAST search on a DNA sequence.

1. For most human genes, there exist related genes. These can be either members of a human gene family, or related genes in other species. To search the GenBank database, go to the NCBI Web site on the BLAST page: http://www.ncbi.nlm.nih.gov:80/BLAST/.

2. Click on "Basic BLAST search."

3. Open your notebook and copy the sequence that you wish to <u>BLAST</u>. This can be either a nucleotide (DNA) sequence, or an amino acid (protein) sequence.

4. If you are starting with a DNA sequence, use the default "blastn" setting and use the "nr" (non-redundant) database. These settings can be changed with the pull-down menus in the middle of the page.

5. Click SEARCH. You will get a page presenting a "request ID" number and an estimate of the time it will take to give you the results of the search.

6. Click the "Format results" button to get the search results.

7. To perform the BLAST search with an amino acid sequence, copy the amino acid sequence of p53 or globin into the box of the BLAST page and choose blastp as the program.

Task 6: <u>Read the BLAST search results.</u>

1. The results of a BLAST search are presented in three formats. At the top of the page is a graphical representation of the search results. The color of the lines tells you how good a match was found. Place the cursor on any line to reveal the name of the sequence that was found to match.

Below the graphical representation is a list of all the matches. They are listed by how good the match is, with the best matches at the top. To the left of this list is the GenBank reference code, followed by the name of the sequence. To the right is a score for the alignment with a corresponding E-value. The E-value is the probability that the alignment could have occurred by chance. The lower the E-value, the more likely that this represents a biologically meaningful alignment.

2. Below this list, each of the alignments is given. Click on any line in the graphical representation or on the score link in the list of sequences to see the relevant alignment.

Task 7: <u>Interpret the BLAST search results.</u>

1. In the list of the BLAST search results, find three human genes that are closely related (but not identical) to the sequence you entered.

2. Highlight each of the sequence alignments, and copy and paste them into your notebook.

3. If you copy them into a word processing program, set the font to Courier 9 point to maintain the formatting of the alignment.

4. Next, find three sequences from other organisms that are closely related to the sequence you entered.

5. Highlight each of the sequence alignments, and copy and paste them into your notebook.

6. Do a BLAST search of the p53 protein sequence against the genome of *Drosophila*. Did you find a similar protein in the *Drosophila* genome?

Task 8: <u>Explore genomic resources on the Web.</u>

1. To access related genomics information, start by clicking on GeneMap'99 http://www.ncbi.nlm.nih.gov:80/genemap/.

2. Next, check out the page titled "Genes and Disease" http://www.ncbi.nlm.nih.gov:80/disease/. On the left side of this page, you can access information on genes associates with specific diseases, such as cancer.

3. Follow the cancer link http://www.ncbi.nlm.nih.gov:80/disease/Cancer.html/. At the top of the page are numbers referring to human chromosomes.

4. Click on one of these to see a diagram of the chromosome and the location of known genes that are implicated in cancer. For example, the p53 gene is located on chromosome 17, as is the BRCA1 gene, mutations in which can lead to breast cancer .

5. On the left side of the cancer page are links to specific cancers and genes implicated in tumorigenesis. Click the "p53 tumor suppressor" link http://www.ncbi.nlm.nih.gov:80/disease/p53.html/.

6. On the left side of this page, you can follow links to articles about p53. You can also search PubMed http://www.ncbi.nlm.nih.gov:80/entrez/query.fcgi?db=PubMed/ for articles on any biomedical subject.

7. Click on LocusLink http://www.ncbi.nlm.nih.gov:80/LocusLink/ where you can find information on the p53 genetic locus, its map information, a list of GenBank sequences, and a list of additional Web resources about p53.

Protein Expression

The procedures in this lab will enable you to:

- **Grow bacteria that contain a subclone.**
- **Extract proteins from the bacteria.**
- **Separate the proteins by SDS polyacrylamide gel electrophoresis.**

PURPOSE

The purpose of this set of experiments is to generate recombinant protein in bacteria.

To express a recombinant protein, you must first insert the DNA that encodes the protein into an expression cloning vector. This can be done in the Subcloning module.

To get started immediately, go to Task 1.

TABLE OF CONTENTS

Background
Introduction
Conceptual and technical overview
Task 1: Select the DNA subclone that will be used to express the protein
Task 2: Place the flasks in the shaker.
Task 3: Spin down the bacteria.
Task 4: Resuspend the bacteria in TE buffer.
Task 5: Add SDS buffer to the bacteria
Task 6: Load the polyacrylamide gel
Task 7: Run the gel.
Task 8: Determine the sizes of the recombinant proteins.
Troubleshooting
Equipment and reagent information

BACKGROUND

In addition to their function in making large amounts of cloned foreign DNA, bacteria can also be used to make foreign proteins. For this technique, signals recognized by the bacteria must be present in the expression vector for correct transcription and translation of the inserted gene. Moreover, because bacteria can not splice out exons, the foreign gene must be in the form of a cDNA. Bacteria cannot perform many of the post-translational modifications such as glycosylation that may be required for a protein's activity. If this type of modification is needed, proteins can be expressed in yeast cells or even in human cell lines.

APPLICATIONS

A powerful application of recombinant DNA technology has been the production of proteins that can be used to treat human diseases. One of the first examples was the production in bacteria of human insulin for the treatment of diabetes.

Recombinant proteins are also used to generate antibodies that recognize the native protein. These serve as important tools in genetics research.

Task 1, 2, 3, 4, 5, 6, 7, 8

INTRODUCTION

To make a recombinant protein in bacteria, you need to grow a large amount of the bacteria that contains your subclone. Expression from the cloning vector is under the control of the regulatory sequences of the lacZ operon.

Protein expression is repressed except when lactose is present in the bacterial growth medium. This feature is useful for expressing recombinant proteins, because many recombinant proteins are either toxic to bacteria, or they slow down their growth.

If the proteins were expressed continuously, it would be difficult or impossible to generate large quantities of bacteria. Not adding lactose to the bacterial will result in no production of recombinant protein until the bacteria have multiplied.

When there are sufficient bacteria containing the subclone, IPTG, a molecule that acts like lactose, is added. This causes the bacteria to start making large amounts of the recombinant protein. Even if the protein is toxic to the bacteria, usually sufficient amounts can be made before the bacteria start to die.

The next step is to collect the bacteria by spinning the culture in a centrifuge. Because the bacteria are have a higher density than the bacterial culture medium, they will collect at the bottom of the centrifuge tube. The medium is then poured off, and the bacteria are suspended in a small amount of TE buffer.

Heat and sodium dodecyl sulfate (SDS), which is a detergent, are used to release the protein from the bacteria. Finally, the proteins are separated by size using polyacrylamide gel electrophoresis (PAGE).

To determine whether the bacteria have made the recombinant protein and to distinguish the recombinant protein from all the other proteins made by the bacteria, you need to perform two side-by-side comparisons.

The first comparison is between the proteins made when induced by IPTG and the proteins from the same bacteria when no IPTG is added. When IPTG is added, there should be an additional protein band in the induced cultures, compared with uninduced cultures in which IPTG has not been added.

The second comparison is between the bacteria that contain only the "empty" expression cloning vector (with no inserted DNA fragment) and bacteria that contain the expression cloning vector with inserted DNA. The induced protein band from the bacteria with the inserted DNA should be larger than the induced protein band from the bacteria with the empty cloning vector. The size difference of this protein band represents the protein you have fused to the maltose binding protein in the expression cloning vector.

Recombinant proteins can be used to generate antibodies that specifically recognize these proteins. The antibodies can then be used to localize the protein of interest in sections of tissue, or to determine the sizes and amounts of the protein present in healthy or disease-carrying individuals.

Task 1, 2, 3, 4, 5, 6, 7, 8

CONCEPTUAL AND TECHNICAL OVERVIEW

Expression Cloning Vector, pMAL

The expression cloning vector used in GDL is the plasmid pMAL. Successful subcloning into pMAL results in a fusion protein, consisting of the *E. coli* maltose binding protein fused to the inserted protein. Click here to view a detailed map of the pMAL plasmid.

To make a fusion protein, the open reading frame (ORF) of the inserted DNA must be in the same reading frame as that of malE, which codes for maltose binding protein. Because of the high affinity of maltose binding protein for maltose, fusion proteins can purified on beads that have maltose attached to them. More information on subcloning into pMAL is available in the Subcloning module.

Blue/White Screening

Several cloning vectors exhibit a change in color of the bacterial colony to indicate that a DNA fragment has been successfully inserted into it. The pMAL expression vector uses modifications of the lactose operon for this purpose. The lactose operon includes a series of bacterial genes that respond to the sugar lactose. One of these genes is lacZ, which codes for beta-galactosidase.

Included in the cloning vector is a truncated portion of lacZ known as lacZ', which encodes the first 146 amino acids of beta-galactosidase. The host bacteria have a mutated form of lacZ which, on its own, is not functional. However, when the cloning vector makes the lacZ' fragment, this can combine with the host bacteria's mutated form of lacZ to make a functional enzyme.

Upstream (in the 5' direction) of the lacZ' coding region are the control sequences of the lacZ operon. These sequences control whether the genes in the operon are expressed, or repressed (not expressed). In the absence of lactose in the bacterial growth medium, expression is repressed. When lactose, or IPTG (an analog of lactose) is added to the medium, the repression is released, and high-level expression occurs.

X-gal (5-bromo-4-chloro-3-indole-beta-D-galactoside) is a substrate of beta-galactosidase. It is normally colorless, but turns blue when cleaved by beta-galactosidase. Thus, when bacteria containing an intact cloning vector are plated on medium containing IPTG and X-gal, lacZ is no longer repressed, and the colonies turn blue.

In the expression vector pMAL the lacZ' coding sequence comes immediately after the malE coding sequence. When the cloning vector is intact (with no inserted DNA) a fusion protein is made in which the first half is maltose binding protein and the second half is the beta galactosidase fragment.

Expression of this protein is controlled by the regulatory sequences of the lacZ operon. Thus, on IPTG and X-gal plates, bacteria that contain intact pMAL are blue. When a DNA fragment is inserted into the pMAL polylinker it disrupts the fusion to lacZ'. Thus, on plates containing IPTG and X-gal, bacteria that contain DNA fragments ligated into pMAL are white.

Task 1, 2, 3, 4, 5, 6, 7, 8

Polyacrylamide Gel Electrophoresis

Polyacrylamide gel electrophoresis (PAGE) is used to separate proteins in an electrical field. Unlike the fairly uniform electrical charges of DNA and RNA, which increase with the size of the molecule, the electrical charges of proteins vary, depending on the charges of the constituent amino acids.

Thus, to use PAGE to separate proteins according to their molecular weight, the proteins must have a uniform charge according to their size. This is done by treating the proteins with the detergent sodium dodecyl sulfate (SDS).

Molecules of SDS bind tightly along the entire length of the protein. The number of SDS molecules bound is proportional to the length of the protein. Because each SDS molecule has a negative charge, SDS gives an overall negative charge to the protein approximately proportional to its molecular weight.

When combined with high temperature, SDS performs a second function, which is to help to denature or pull apart the bonds that allow a protein to maintain its shape. When proteins are denatured, they tend to behave in a similar manner when being pulled through the pores of an acrylamide gel.

The use of SDS in polyacrylamide gel electrophoresis is referred to as SDS-PAGE.

To determine the sizes of proteins run on SDS-PAGE, marker proteins of known size are run in a lane of the gel. The positions of proteins in other lanes of the gel are compared with the marker proteins to determine approximate sizes (Task 8). Because SDS does not completely denature proteins, size determinations by SDS-PAGE are not as accurate as size determinations of DNA fragments made with agarose gel electrophoresis.

Task 1, 2, 3, 4, 5, 6, 7, 8

PROCEDURES

Task 1: <u>Select the DNA subclone that will be used to express the protein.</u>

1. Click the arrow on the petri dish to get the DNA subclone selection menu.

2. Select a plate with bacterial colonies that you produced in the Subcloning module. Alternatively, you can select one of the "test" DNA subclones. The colonies will now appear in the petri dish on the bench.

3. Drag a toothpick from the beaker to the petri dish.

4. Using the toothpick, select the colony that contains the construct from which you wish to make protein. <u>Select a colony</u> by releasing the mouse when the end of the toothpick is over the colony you want.

5. Drag the toothpick, which now has some bacteria on it, to one of the flasks on the bench. The flasks contain 50 milliliters of rich bacterial medium known as <u>Luria broth</u>.

6. Release the toothpick over the mouth of the flask. The toothpick will drop into the flask.

7. Because you will want to compare the results of inducing protein expression to not inducing expression, inoculate a second white colony from the same petri dish into another flask.

8. In order to compare proteins made from bacteria that contain only the expression cloning vector, inoculate another flask with a blue colony.

9. If you have picked up a colony you don't want, drag the toothpick to the trash can and release it.

Task 2: <u>Place the flasks in the shaker.</u>

1. When you have inoculated all the flasks you want to use, drag a cap onto each flask.

2. Drag the flasks, one by one, to the <u>shaker</u>.

3. Release each flask over a holder in the shaker; it will snap into place.

4. When all the flasks are in the shaker, click on the top of the shaker to close it.

5. Turn the shaker on by clicking the ON/OFF switch. The shaker has been set to 37 degrees C, the optimal temperature for growth of *E. coli* bacteria.

6. Set the <u>timer</u> for 8 hours. This will allow sufficient time for the bacteria to divide and multiply before you induce expression of the recombinant protein.

7. When the incubation is complete, click the ON/OFF switch to stop the shaker, then open the shaker top by clicking on it.

8. Remove the tops of the flasks in which you want to induce protein expression.

9. Use the <u>pipettor</u> to add 60 microliters of <u>IPTG</u>, which is found in a tube in the small tube rack, to the flasks in which you wish to induce protein expression. Caution: Be sure you do not add IPTG to the flasks you are using as your uninduced control.

10. Put the tops back on the flasks, close the shaker top by clicking on it, and turn the shaker on.

11. Set the timer for an additional 8 hours, to allow the bacteria to express the recombinant protein. <u>Click here to view an animation of how the bacteria produce recombinant protein.</u>

Task 3: <u>Spin down the bacteria.</u>

1. When the incubation is complete, stop and open the shaker.

2. Drag each flask to the bench near the rack that contains centrifuge tubes.

3. Pour the contents of each flask into a centrifuge tube by dragging the flask over a centrifuge tube.

4. When the contents of all the flasks have been poured into the centrifuge tubes, drag the caps of the centrifuge tubes onto the tubes.

5. Drag the centrifuge tube rack with the capped centrifuge tubes to the next screen.

6. Drag the centrifuge tubes, one by one into the <u>centrifuge</u>. Release the tube over a hole in the centrifuge holder; the tube will drop into the holder.

7. When all the tubes are in the centrifuge, click on the centrifuge top to close it.

8. Set the speed of the centrifuge by clicking on the speed display window. On the slider set the speed to 5000 rpm.

9. Click the time display on the centrifuge and set the time to 10 minutes.

10. Turn on the centrifuge by clicking the ON/OFF switch.

Task 4: <u>Resuspend the bacteria in buffer.</u>

1. After the centrifuge run is complete, click the centrifuge door to open it.

2. Drag each tube to the centrifuge tube rack.

3. Remove the caps from the tubes by dragging them to the bench.

4. Drag each tube to the waste jar and pour off the liquid. You should see a pellet of bacteria at the bottom of the tube.

5. When you have poured off the liquid from all the tubes, use the pipettor to add 1 milliliter of <u>TE buffer</u> to each of the tubes. The TE buffer is found in the small tube rack.

6. Drag each of the tubes to the <u>vortex mixer</u>. You do not have to have the caps on the tubes for this step. To activate the vortex mixer, position the tube so that its base touches the black padded surface of the mixer. The rapid agitation causes the bacterial cells to be pulled off of the side of the centrifuge tube and become suspended in the TE buffer.

Task 5: <u>Add SDS buffer to the bacteria</u>.

1. When all the bacteria have been resuspended, drag the centrifuge tube rack with the centrifuge tubes to the next screen.

2. Use the pipettor to transfer 50 microliters of the resuspended bacteria from the centrifuge tubes to the microcentrifuge tubes. Caution: Be sure not to mix any of the bacteria from one tube with the bacteria from another tube.

3. Use the pipettor to add 50 microliters of <u>SDS buffer</u> to each of the microcentrifuge tubes that have bacteria in them.

4. After adding the SDS, drag each of the microcentrifuge tubes to the <u>heat block</u>.

5. Turn the heat block on and set the temperature to 98 degrees C. Set the timer for 5 minutes.

6. When the incubation is complete, drag the tubes back to their rack.

Task 6: Load the polyacrylamide gel.

1. You will now use polyacrylamide gel electrophoresis (PAGE) to separate the resulting DNA fragments. Click here to view a movie on how to prepare a polyacrylamide gel.

2. Use the pipettor to transfer 10 microliters of the marker proteins found in the small rack to at least one of the gel wells. The marker is usually loaded in the first well on the left.

3. When the pipettor tip is over the selected well, release its contents into the well by releasing the mouse button.

4. Use the pipettor to transfer 10 microliters of the bacterial protein and SDS buffer from each of the tubes to a different well of the gel.

5. Load all the samples into the wells.

Task 7: Run the gel.

1. When all the wells are loaded, drag the cover onto the gel box. It is correctly positioned when you hear it click into place.

2. Adjust the current by clicking on the voltage/amperage indicator. 15 milliAmperes (mA) is an appropriate current for this gel.

3. Turn on the power supply by clicking the ON/OFF switch. A new window will open to give you a good view of the gel inside the gel box. You will see the proteins as they would appear when stained with Coomassie blue stain.

4. Use the slider to adjust the current while the gel is running.

5. Click STOP to halt the electrophoresis. The goal is to get maximum separation of the proteins, particularly those in the size range of your recombinant protein.

6. Click on the camera icon to take a picture of the gel. The picture will be entered into your notebook. You can take up to three pictures of each gel.

7. To return to the lab bench, close the window.

Task 8: Determine the sizes of the recombinant proteins.

1. From your notebook, find the pictures you took of the gel. Open the one that shows the best separation of the proteins.

2. Compare the pattern found in the lane made from bacteria to which you added IPTG with the one from bacteria to which you did not add IPTG. If the subclone in these bacteria contained the pMAL expression vector, you should see a dark band made by the induced bacteria which is not made by the uninduced bacteria. If you do not see this difference, check the troubleshooting section to determine what might have gone wrong.

3. Next, compare the pattern of proteins made from induced bacteria containing pMAL alone with the pattern from bacteria that had the pMAL vector with a DNA fragment inserted into it. If the DNA fragment coded for a protein (globin or p53) and it was inserted in the correct reading frame, you should see a dark band that is higher in the lane from these bacteria than the band from the bacteria that contained pMAL alone.

4. To estimate the size of the recombinant protein, compare its position on the gel with the position of the marker proteins. Those that appear at about the same place have about the same size.

5. Making an accurate determination of the sizes of proteins separated by PAGE is very similar to determining the sizes of DNA molecules separated by agarose gel electrophoresis, as performed in the Restriction Mapping module. Using the ruler, measure the distance from the starting well to each of the <u>marker proteins</u>, and plot this information on a piece of graph paper or in a graphing program.

6. Make the Y axis the logarithm of each protein's molecular weight (in Daltons or kiloDaltons) and the X axis the distance migrated.

7. Draw a straight line through the plotted points.

8. Determine the distance each of the induced proteins migrated from the starting well, from the culture that contained pMAL alone and the one that contained pMAL with a DNA fragment inserted into it.

9. Plot each of these distances on the graph you made of the marker protein sizes versus distance, and determine the approximate molecular weights of the induced proteins.

10. Subtract the size of maltose binding protein (approximately 40,000 Daltons) from the size you calculated of the induced protein made from your subclone.

11. Calculate the size of the protein you would predict from the DNA sequence you inserted into the pMAL vector. This is done by dividing the number of bases in the DNA fragment by 3. This will give you the number of amino acids in the protein made from this DNA (because each codon is made of three bases).

12. Multiply the number of amino acids by 100 Daltons, an average of the molecular weight for an amino acid. If you want to be more precise, you can look up the exact molecular weight for each amino acid found in the protein the DNA would encode and add these together to get the total molecular weight.

13. If the molecular weight of the induced protein made from your subclone is equal to the molecular weight of maltose binding protein plus the molecular weight you calculated from the DNA sequence, you have successfully produced recombinant protein. If this was not the case, check the <u>troubleshooting section</u> to determine what might have gone wrong.

<u>TROUBLESHOOTING</u>

If no protein appears in a lane, the most likely causes are:
- The bacterial cultures were not centrifuged long enough or fast enough.
- You forgot to pour off the liquid from the bacterial pellet after centrifugation.
- You forgot to resuspend the bacterial pellet with the vortex mixer.
- You forgot to add the SDS buffer to the resuspended bacteria in the microcentrifuge tube.
- You didn't heat the SDS and bacterial mixture in the heat block for the right amount of time, or the heat block wasn't set to the right temperature.
- You forgot to load the protein onto the gel.

If there was no difference between the proteins from the uninduced bacterial culture and the cultures that should have been induced, the most likely causes are:
- You forgot to add IPTG to the cultures you wanted to induce.
- The colony you selected from the petri dish didn't contain a protein expression vector.

If there was no difference between the proteins from the culture that should have made recombinant protein and the culture with the protein expression vector alone, the most likely cause is:
- You picked up a blue colony instead of a white one for subcloning plate.
- The inserted fragment in the subclone was not in the right reading frame.

If there was very little recombinant protein, the most likely cause is:
- You didn't leave the flasks in the shaker long enough after you added IPTG.
- You didn't add enough IPTG.

EQUIPMENT AND REAGENT INFORMATION

Pipettor

Drag the automatic pipettor from its rack and a new disposable tip will appear. The tip will be automatically discarded after each use.

Click on the dark rectangular window on the pipettor to get a pop-up slider. Select the volume by adjusting the slider, by using either the sliding needle or the arrow buttons for fine control. Click OK or anywhere outside the slider panel to dismiss it.

Drag the pipettor to the rack containing tubes. When the tip of the pipettor is over a tube, the contents of the tube will be visible. This tells you which tube the pipettor tip is in.

When you release the mouse, if the volume is set within the correct limits, the reagent will be drawn up into the pipettor tip. If not, you will get an error message and will need to reset the volume.

Drag the pipettor to the tube to which you want to add the reagent. When the pipettor tip is over the tube, release the mouse. What was in the pipettor tip will now be added to that tube.

Shaker

The shaker is used to grow bacterial cultures. Its temperature is set at 37degrees C, which is the ideal temperature for growing bacteria such as the common lab strains of *E. coli*.

For optimal growth of aerobic bacteria such as *E. coli*, the culture must be constantly mixed with air. This is achieved by rapid shaking of the bacteria in specially designed flasks that have indentations along their base. These promote mixing and air flow in the bacterial culture.

To close the shaker, click on its top. Then click the ON/OFF switch to turn it on. The shaker must be closed to turn it on.

TE Buffer

The commonly used laboratory reagent known as TE buffer is made up of two ingredients: the relatively inexpensive buffering agent Tris HCl and the chelating agent EDTA.

EDTA is added because most DNA degrading enzymes (DNases) require magnesium ions to function. EDTA binds, or chelates, magnesium ions, thereby reducing the likelihood that DNases will destroy the DNA.

Vortex Mixer

Vortex mixers are used to rapidly mix solutions in test tubes. Most vortex mixers have a touch-sensitive surface, so that placing a tube on the surface activates it. When activated, the vortexer surface vibrates rapidly, causing the contents of the tube that is in contact with it to be shaken and thoroughly mixed.

Centrifuge

A centrifuge is used to separate biological materials based on their density or mass. Two parameters determine how well separated the materials are: the speed and duration of the spinning of the materials in the centrifuge.

Centrifuge speed depends on how different the densities or masses of the two materials are, and the design of the centrifuge. In general, the greater the density or mass difference, the easier it is to separate them. Thus, they require less speed and less time to separate.

Centrifuge design is important, because separating the materials depends on the force that can be exerted on them, measured in multiples of the force of gravity (g). A centrifuge's instruction book contains a conversion table for the speed at which the rotor turns and the force generated on tubes in the rotor. The force also varies with the type of rotor. Once you know the force exerted on the sample, you can calculate the amount of time required to get good separation of two samples of known mass.

Separating bacteria from culture media is a fairly simple process, since the density differences are quite large. For the centrifuge used in this module, a speed of 5000 rpm for 10 minutes is sufficient to bring most of the bacteria to the bottom of the tube.

Heat block

Turn on the heat block by clicking the ON/OFF switch.

Click the control knob on the heater. A slider appears that will allow you to set the temperature.

Timer

When you click on the timer a slider will appear. Set the slider on the timer for the desired time. Click the GO button on the timer to start the timer. In GDL, time is speeded up, so the timed process will take only a few seconds.

If you need to adjust anything, click outside the time slider to dismiss it without starting the timer.

Luria Broth

One liter of Luria broth (LB medium or Luria-Bertani medium) is made by adding to 950 ml of water: 10 grams of bacto-tryptone, 5 grams of yeast extract and 10 grams of NaCl. The mixture is shaken then the pH is adjusted to 7.0. To sterilize the mixture it is autoclaved for 20 minutes.

IPTG

IPTG (isopropyl beta-D-thiogalactopyranoside) is an analog of lactose. When added to culture media, bacteria will take it up, and its effect will be similar to the addition of lactose to the media. However, IPTG is not metabolized (broken down) by bacteria. Thus, it can act as a more potent inducer than lactose, which bacteria use for food.

A concentrated stock solution of IPTG is usually made. In GDL, a 250-mM stock solution is used. The final concentration of IPTG in the bacterial culture should be approximately 0.3 mM.

Thus, for a 50-milliliter culture of bacteria, you need to add 60 microliters of the stock solution of IPTG.

SDS Buffer

SDS buffer consists of the following ingredients: 100 mM Tris-HCl (pH 6.8), 4% weight/volume (w/v) sodium dodecyl sulfate (SDS), 0.2% w/v bromophenol blue, and 20% w/v glycerol.

The Tris-HCl buffers the acidity of the solution and provides ions for electrophoresis. Molecules of SDS (a detergent) bind tightly along the entire length of the protein. The number of SDS molecules bound is proportional to the length of the protein. Because each SDS molecule has a negative charge, SDS gives an overall negative charge to the protein approximately proportional to its molecular weight.

When combined with high temperature, SDS performs a second function, which is to help to denature or pull apart the bonds that allow a protein to maintain its shape.

Bromophenol blue is a dye that lets you determine how far the proteins have run in the gel. Glycerol causes the solution to sink to the bottom of the gel well. Just before use, a reducing agent such as dithiothreitol (DTT) is usually added to the buffer. This helps ensure that the protein is denatured in the gel.

Marker Proteins

Marker proteins are proteins of known molecular weight that can be used to determine the size of other proteins separated by polyacrylamide gel electrophoresis (PAGE).

The marker proteins used in GDL and their molecular weights (in Daltons) are: myosin – 200,000; beta-galactosidase – 116,250; phosphorylase b – 97,400; bovine serum albumin – 66,200; ovalbumin – 45,000; carbonic anhydrase – 31,000; soybean trypsin inhibitor – 21,500; lysozyme – 14,400; aprotinin – 6,500.

Coomassie Blue Stain

Coomassie blue is a dye that binds to proteins primarily through interactions with basic and aromatic amino acids. The amount of bound Coomassie dye more or less depends on the amount of protein present. Therefore, how dark a protein band in a gel is after staining with Coomassie blue is a fairly good indication of the abundance of that protein.

Normally, staining with Coomassie blue dye is done after the polyacrylamide gel run has been completed. In GDL, the proteins in the gel can be seen as if they have already been stained with Coomassie blue dye.

Notebook

This is where you record your observations.

To review any of the photographs you've taken, click the link in your notebook to make the viewing window appear.

You can also use the BACK and NEXT buttons in the viewing window to navigate through your photograph collection.